Visible Light Communication and Positioning

Visible Light Communication and Positioning

Special Issue Editor

Chen Gong

MDPI • Basel • Beijing • Wuhan • Barcelona • Belgrade

MDPI

Special Issue Editor
Chen Gong
University of Science and
Technology of China
China

Editorial Office
MDPI
St. Alban-Anlage 66
4052 Basel, Switzerland

This is a reprint of articles from the Special Issue published online in the open access journal *Electronics* (ISSN 2079-9292) from 2018 to 2019 (available at: https://www.mdpi.com/journal/electronics/special_issues/Visible_Light).

For citation purposes, cite each article independently as indicated on the article page online and as indicated below:

LastName, A.A.; LastName, B.B.; LastName, C.C. Article Title. *Journal Name* **Year**, *Article Number, Page Range.*

ISBN 978-3-03921-435-8 (Pbk)
ISBN 978-3-03921-436-5 (PDF)

Contents

About the Special Issue Editor

Chen Gong received the B.S. degree in electrical engineering and mathematics (minor) from Shanghai Jiaotong University, Shanghai, China, in 2005 and the M.S. degree in electrical engineering from Tsinghua University, Beijing, China, in 2008; and the Ph.D. degree from Columbia University, New York City, NY, USA, in 2012. He was a Senior Systems Engineer with the Qualcomm Research, San Diego, CA, USA, from 2012 to 2013. He is currently a Faculty member with the University of Science and Technology of China. His research interests are in wireless communications, optical wireless communications, and signal processing. He was selected by the Young 1000 Talent Program of China Government in 2014, and awarded by Hongkong Qiushi Outstanding Young Researcher Award in 2016.

electronics

MDPI

Editorial

Visible Light Communication and Positioning: Present and Future

Chen Gong

Department of Electronic and Information Science, University of Science and Technology of China, Hefei 230027, Anhui, China; cgong821@ustc.edu.cn

Received: 25 June 2019; Accepted: 11 July 2019; Published: 15 July 2019

1. Introduction

Future wireless communication may extend its spectrum to visible light due to its potential large bandwidth. It serves as a promising candidate for high-speed, line-of-sight communication. Besides, due to its lack of electromagnetic radiation and immunity to electromagnetic interference, the visible light spectrum can be deployed for the industrial Internet of Things. Its limited transmission range can alleviate the interference issue and can lead to ultra-dense transmitter and receiver deployment. Current research into visible light communication includes the experimental demonstration of high-speed communication systems [1,2], beamforming optimization [3], the physical-layer secrecy problem [4], and multi-user coverage [5].

Besides communication, the limited transmission range can lead to high positioning accuracy, especially for indoor visible light positioning (VLP). The received signal strength (RSS)-based VLP using photodiode and the angle of arrival (AOA)-based VLP using camera are two mainstream approaches. While the former approach can achieve a positioning accuracy of several centimeters, the latter one can achieve a positioning accuracy within one centimeter. A summary of current progress on indoor visible light positioning is shown in the Table 1.

Table 1. Summary of current progress on indoor visible light positioning. RSS: received signal strength; AOA: angle of arrival.

Ref.	Algorithm	Accuracy (cm)	Number of TX LEDs	Receiver Realization	LED Height (cm)	Note
[6]	RSS	2.4	3	Single PD	60	
[7]		1.66	3		100	Compensation of Positioning Error
[8]	Finger Print	5	2		167	Image Sensor Acceleration
[9]	AOA	1.53	4		72	Error Cancellation
[10]		6.6	3		180	
[11]	SVD	6	3	Camera	120	
[12]	Bayesian	0.86	4		190	Industrial Camera, Optical Compensation
[13]	Differential	4	3		100	Differential Detection
[14]	Image Processing	<10	24		300	Fisheye Camera
[15]		4.81	4		50	
[16]		1	3		231	Shift and Rotation based on a Reference Point
[17]	Differential AOA	<6	4		113	Unknown Tilting Angle

For a more comprehensive overview of visible light communication and positioning, the readers may refer to [18,19], respectively.

2. The Present Issue

The present issue, named "Visible Light Communication and Positioning", focuses on visible light communication and visible light positioning, in which four papers explore visible light communication and three papers investigate visible light positioning.

For visible light communication, the published works focus on the devices, the physical-layer techniques, and system work aspects. In [20], the light-to-frequency converter for VLC is characterized. In [21,22], the physical-layer non-orthogonal multiple access and multi-color VLC, respectively, are addressed. In [23], the system-level VLC based on the software-defined radio with intelligent transportation and indoor applications is addressed.

Besides VLC, in [24–26], visible light positioning is explored. A fingerprint-based indoor positioning system for multiple reflections is proposed in [24]. To address the issue of non-perfect LED deployment, in [25], the impact of LED tilt on visible light positioning accuracy is analyzed. Moreover, a mobile optoelectronic tracking system based on feedforward control is investigated in [26].

3. Future

While this special issue focuses on visible light communication and visible light positioning, more fundamental works into joint performance optimization need future work. For example, the impact of LED layout on the communication performance and the positioning accuracy, as well as the related joint optimization for both communication and positioning, remain to be investigated.

Acknowledgments: First of all, we would like to thank all researchers who submitted articles to this special issue for their excellent contributions. We are also grateful to all reviewers who helped in the evaluation of the manuscripts and made very valuable suggestions to improve the quality of contributions. We would like to acknowledge the editorial board of *Electronics*, who invited us to guest edit this special issue. We are also grateful to the *Electronics* Editorial Office staff who worked thoroughly to maintain the rigorous peer-review schedule and timely publication.

Conflicts of Interest: The author declares no conflicts of interest.

References

1. Werfli, K.; Chvojka, P.; Ghassemlooy, Z.; Hassan, N.B.; Zvanovec, S.; Burton, A.; Haigh, P.A.; Bhatnagar, M.R. Experimental Demonstration of High-Speed 4 × 4 Imaging Multi-CAP MIMO Visible Light Communications. *IEEE J. Lightwave Technol.* **2018**, *36*, 1944–1951. [CrossRef]
2. Bian, R.; Tavakkolnia, I.; Haas, H. 15.73 Gb/s Visible Light Communication With Off-the-Shelf LEDs. *IEEE J. Lightwave Technol.* **2019**, *37*, 2418–2424. [CrossRef]
3. Ling, X.; Wang, J.; Liang, X.; Ding, Z.; Zhao, C.; Gao, X. Biased Multi-LED Beamforming for Multicarrier Visible Light Communications. *IEEE J. Sel. Areas Commun.* **2018**, *36*, 106–120. [CrossRef]
4. Wang, J.; Liu, C.; Wang, J.; Wu, Y.; Lin, M.; Cheng, J. Physical-layer Security for Indoor Visible Light Communications: Secrecy Capacity Analysis. *IEEE Trans. Commun.* **2018**, *66*, 6423–6436. [CrossRef]
5. Yin, L.; Haas, H. Coverage Analysis of Multiuser Visible Light Communication Networks. *IEEE Trans. Wirel. Commun.* **2018**, *17*, 1630–1643. [CrossRef]
6. Kim, H.S.; Kim, D.R.; Yang, S.H.; Son, Y.H.; Han, S.K. An indoor visible light communication positioning system using a RF carrier allocation technique. *IEEE J. Lightwave Technol.* **2012**, *31*, 134–144. [CrossRef]
7. Jeong, E.M.; Yang, S.H.; Kim, H.S.; Han, S.K. Tilted receiver angle error compensated indoor positioning system based on visible light communication. *Electron. Lett.* **2013**, *49*, 890–892. [CrossRef]
8. Tanaka, T.; Haruyama, S. New position detection method using image sensor and visible light leds. In Proceedings of the IEEE International Conference on Machine Vision, Dubai, UAE, 28–30 December 2009.

9. Pan, W.; Hou, Y.; Xiao, S. Visible light indoor positioning based on camera with specular reflection cancellation. In Proceedings of the IEEE Conference on Lasers and Electro-Optics Pacific Rim (CLEO-PR), Singapore, 31 July–4 August 2017.
10. Lin, B.; Ghassemlooy, Z.; Lin, C.; Tang, X.; Li, Y.; Zhang, S. An indoor visible light positioning system based on optical camera communications. *IEEE Photonics Technol. Lett.* **2017**, *29*, 579–582. [CrossRef]
11. Zhang, R.; Zhong, W.D.; Qian, K.; Wu, D. Image sensor based visible light positioning system with improved positioning algorithm. *IEEE Access* **2017**, *5*, 6087–6094. [CrossRef]
12. Guan, W.; Chen, X.; Huang, M.; Liu, Z.; Wu, Y.; Chen, Y. High-speed robust dynamic positioning and tracking method based on visual visible light communication using optical flow detection and bayesian forecast. *IEEE Photonics J.* **2018**, *10*, 1–22. [CrossRef]
13. Lv, H.; Feng, L.; Yang, A.; Guo, P.; Huang, H.; Chen, S. High accuracy VLC indoor positioning system with differential detection. *IEEE Photonics J.* **2017**, *9*, 1–13. [CrossRef]
14. Nakazawa, Y.; Makino, H.; Nishimori, K.; Wakatsuki, D.; Kobayashi, M.; Komagata, H. High-speed, fish-eye lens equipped camera based indoor positioning using visible light communication. In Proceedings of the 2015 International Conference on Indoor Positioning and Indoor Navigation (IPIN), Banff, AB, Canada, 13–16 October 2015.
15. Li, Y.; Ghassemlooy, Z.; Tang, X.; Lin, B.; Zhang, Y. VLC smartphone camera based indoor positioning system. *IEEE Photonics Technol. Lett.* **2018**, *30*, 1171–1174. [CrossRef]
16. Xu, J.; Gong, C.; Xu, Z. Indoor Visible Light Positioning with Centimeter Accuracy Based on A Commercial Smartphone Camera. *IEEE Photonics J.* **2018**, *10*, 1–6.
17. Zhu, B.; Cheng, J.; Yan, J.; Wang, J.; Wang, Y. VLC positioning using cameras with unknown tilting angles. In Proceedings of the GLOBECOM 2017—2017 IEEE Global Communications Conference, Singapore, 4–8 December 2017.
18. Yassin, A.; Nasser, Y.; Awad, M.; Al-Dubai, A.; Liu, R.; Yuen, C.; Raulefs, R.; Aboutanios, E. Recent Advances in Indoor Localization: A Survey on Theoretical Approaches and Applications. *IEEE Commun. Surv. Tutor.* **2016**, *19*, 1327–1346. [CrossRef]
19. Hassan, N.; Naeem, A.; Pasha, M.; Jadoon, T.; Yuen, C. Indoor positioning using visible led lights: A survey. *ACM Comput. Surv.* **2015**. [CrossRef]
20. Martínez Ciro, R.; López Giraldo, F.; Betancur Perez, A.; Luna Rivera, M. Characterization of Light-To-Frequency Converter for Visible Light Communication Systems. *Electronics* **2018**, *7*, 165. [CrossRef]
21. Dong, Z.; Shang, T.; Li, Q.; Tang, T. Adaptive Power Allocation Scheme for Mobile NOMA Visible Light Communication System. *Electronics* **2019**, *8*, 381. [CrossRef]
22. Kwon, T.H.; Kim, J.E.; Kim, Y.H.; Kim, K.D. Color-Independent Visible Light Communications Based on Color Space: State of the Art and Potentials. *Electronics* **2018**, *7*, 190. [CrossRef]
23. Martinek, R.; Danys, L.; Jaros, R. Visible Light Communication System Based on Software Defined Radio: Performance Study of Intelligent Transportation and Indoor Applications. *Electronics* **2019**, *8*, 433. [CrossRef]
24. Tran, H.; Ha, C. Fingerprint-Based Indoor Positioning System Using Visible Light Communication—A Novel Method for Multipath Reflections. *Electronics* **2019**, *8*, 63. [CrossRef]
25. Plets, D.; Bastiaens, S.; Martens, L.; Joseph, W. An Analysis of the Impact of LED Tilt on Visible Light Positioning Accuracy. *Electronics* **2019**, *8*, 389. [CrossRef]
26. Luo, Y.; Ren, W.; Huang, Y.; He, Q.; Wu, Q.; Zhou, X.; Mao, Y. Feedforward Control Based on Error and Disturbance Observation for the CCD and Fiber-Optic Gyroscope-Based Mobile Optoelectronic Tracking System. *Electronics* **2018**, *7*, 223. [CrossRef]

electronics

MDPI

Article

Characterization of Light-To-Frequency Converter for Visible Light Communication Systems

Roger Alexander Martínez Ciro [1,*], Francisco Eugenio López Giraldo [1],
Andrés Felipe Betancur Perez [2] and Martín Luna Rivera [3]

[1]   Facultad de Ingenierías, Instituto Tectonológico Metropolitano (ITM), Calle 54A No. 30-01,
     Barrio Boston, CP 050012 Medellín, Colombia; franciscolopez@itm.edu.co
[2]   Departamento de Tecnología Electrónica, Universidad Carlos III de Madrid, Calle de Butarque 15,
     CP 28911 Leganés, Madrid, Spain; abetancu@ing.uc3m.es
[3]   Facultad de Ciencias, Universidad Autónoma de San Luis Potosí (UASLP), Zona Universitaria,
     Av. Salvador Nava s/n, CP 78290 San Luis Potosí, Mexico; mlr@fciencias.uaslp.mx
*   Correspondence: rogermartinez@itm.edu.co; Tel.: +57-4460-0727

Received: 17 July 2018; Accepted: 17 August 2018; Published: 28 August 2018

Abstract: PIN (positive intrinsic negative) photodiodes and analog-to-digital converters (ADC) are commonly used on visible light communication (VLC) receivers in order to retrieve the data on detected signals. In this paper, a visible light communication receiver based on a light to frequency converter (LTF) is proposed. We characterized the LTF and derived an equation for signal-to-noise ratio (SNR) estimation in terms of its input optical power, and the frequency of the output periodic signal. The experiments show that the periodic signal of the LTF converter has a maximum output frequency of 600 kHz at a distance of 6.2 cm. In this setup, measured SNR reached 18.75 dB, while the lowest obtained SNR with 1.1 m length was roughly −35.1 dB. The results obtained suggest that a bit rate of 150 kbps can be achieved with an on-off keying (OOK) modulation format. We analyzed the results and discuss the advantages and limitations of the LTF converter for optical wireless communication purposes.

Keywords: visible light communication; light to frequency converter; white-light LED; optical wireless communication

---

1. Introduction

In optical wireless communication (OWC), it is common to find optical receivers conformed by positive intrinsic negative (PIN) photodiodes or avalanche photodiode (APD) and analog-to-digital converters (ADC) [1–6], as proposed by the IEEE 802.15.7 standard [7] for visible light communication (VLC) systems. Among these, the PIN photodiodes (PD) are presented more often in VLC systems, because they have better immunity to noise and low parasitic capacitance, and can be used to speed up the transmission (ultrafast PIN-PD), which is a milestone in VLC works [2,6]. This is a result of the intrinsic material between the p–n junction, which leads to a reduction of the time constant, and thus better bandwidth [2,8]. In fact, the PIN-type photodiodes adapted to the red-green-blue (RGB) sensors have been studied for applications in VLC systems [9], the authors characterized these sensors and determined its frequency response. Furthermore, PIN photodiodes are being adapted to other systems, to perceive intensity levels of light and turn over periodic electrical signals with frequencies that correspond to the incident power in the photodetector [10]. These devices are known as light-to-frequency (LTF) converters, and are internally comprised of PIN-type photodiodes and a module that transforms the photocurrent to frequency. The mentioned module employs a voltage-controlled oscillator (VCO), which reduces and simplifies the conditioning circuit of the photodiode, allowing its adaptation with embedded low-cost systems that do not require

ADC [10–16]. Therefore, the LTF converters are exposed as an attractive solution for the detection of communication signals in the visible range of the electromagnetic spectrum because of the reduction of the system complexity [10]. Recently, LTF converters have been investigated, for example, in optical communication systems for the design of portable transceivers [11] and in health applications to detect levels of oxygen in the blood [12], among others.

In this paper, we propose the use of a light-to-frequency converter as an alternative to design the receiver of a VLC system. In this scheme, the characterization and performance evaluation of LTF converter as a receiver in a VLC system, based on on-off keying (OOK) modulation, is presented. The main contribution of this article is summarized as follows: initially, the characterization of an LTF converter is presented, and an equation is derived based on both the incident optical power and the frequency generated in the LTF, in order to estimate the system's SNR value. The second contribution is the evaluation of the LTF by using a periodic optical signal, which reveals the advantages and disadvantages regarding its use as a receiver in a VLC system.

The rest of the paper is organized as follows: Section 2 shows the model of the VLC system and the LTF converter. The characterization of the LTF converter for a VLC system is presented in Section 3. The results and discussions are presented in Section 4. Finally, we summarize the main conclusions.

2. VLC and LTF System Model

2.1. VLC System Model

VLC systems are based on intensity modulation and direct detection (IM/DD), which is the most used method to implement optical wireless communications [1,2,7]. A typical VLC system is depicted in Figure 1.

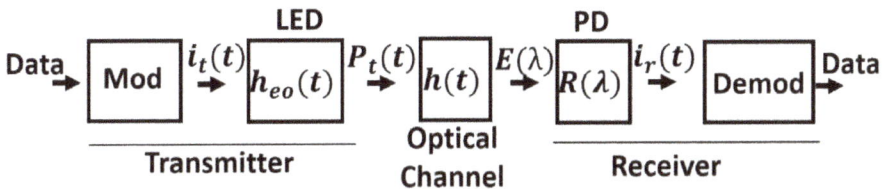

Figure 1. Block diagram of a visible light communication (VLC) system. LED—light-emitting-diode; PD—photodiode.

Once the photodiode (PD) detects an incident optical power or irradiance $E(\lambda)$ on its photosensitive surface, it will generate a photocurrent $i_r(t)$ proportional to device responsivity $R(\lambda)$ and $E(\lambda)$, which is corrupted by noise $n(t)$. Such relation is illustrated in Equation (1):

$$i_r(t) = R(\lambda)E(\lambda) + n(t) \tag{1}$$

If emitted optical signal degradation effects due to communication channel are considered, the model in the work of [1] can be described through the expressions in Equation (2):

$$\begin{aligned} i_r(t) &= R(\lambda)P_t(t) \otimes h(t) + n(t), \\ P_t(t) &= i_t(t) \otimes h_{eo}(t), \end{aligned} \tag{2}$$

where $i_t(t)$ is the bias current of light-emitting-diode (LED); $h_{eo}(t)$ is the impulse response of LED; $P_t(t)$ is the emitted instantaneous optical power by the LED; $h(t)$ represents the channel impulse response; $i_r(t)$ is the sensor generated photocurrent; $R(\lambda)$ is the photodiode responsivity; $\otimes$ denotes the convolution operator; and $n(t)$ is the system noise, which is modeled as additive white gaussian noise (AWGN). The radiated optical power is always positive $P_t(t) \geq 0$; moreover, it is important take into account that the required illumination in a space in which people are dwelling needs to be below a certain limit of the average total emitted optical power, in order to mitigate the possible harmful effects on the eyes [2]. The average optical power of the source can be estimated with Equation (3):

$$P_{avg} = \lim_{T \to \infty} \frac{1}{2T} \int_{-T}^{T} P_t(t)dt \tag{3}$$

VLC systems usually have two main threats: thermal noise and shot noise, both of which distort the signal of interest. The source of shot noise is the randomness in the photon absorption process and the electron-hole pair recombination within PD, whereas thermal noise depends on the environment temperature that perturbs enough the electrons in the receiver discrete devices [8]. The noise is a random process, thus it is characterized by a total variance. In the model described in the literature [8], the overall variance σ^2 is equal to the sum of the shot noise variance σ^2_{shot} and the thermal noise variance $\sigma^2_{thermal}$ as shown in Equation (4):

$$\sigma^2 = \sigma^2_{shot} + \sigma^2_{thermal},$$
$$\sigma^2 = 2qR(\lambda)(P_r + P_n)B_w + i_{amp}{}^2 B_{amp}, \tag{4}$$
$$B_n = \beta B_r,$$

where:

q, is the electron charge $(1602 \times 10^{-19}$ coulomb$)$,
$R(\lambda)$, is the PD responsivity,
P_r is the signal power received,
P_n is the noise power generated by external light sources,
B_w is the channel equivalent noise bandwidth,
i_{amp} is the parasitic current of the amplifier,
B_{amp} is the bandwidth of the amplifier,
β is the Bandwidth factor, and
B_r is the signal bit rate.

A commonly used figure of merit in telecommunications is the SNR, which is a ratio between the signal power and the power contributed by the noise described by σ^2 [17–23]. In the case of a VLC system, the electrical SNR can be estimated with (5):

$$P_{in}(t) = E(\lambda)Ar,$$
$$SNR = \frac{(R(\lambda),P_{in}(t))^2}{\sigma^2} \tag{5}$$

where:

$P_{in}(t)$ is the incident optical power,
$E(\lambda)$ is the irradiance, and
Ar, is the PD area.

2.2. Light-To-Frequency Model

An LTF reduces and simplifies the signal acquisition process coming from a light source because its output can be sent directly to be processed to a microcontroller for data processing [10]. As a result, traditional systems using ADCs can be seen as an additional option on the list. In some low-cost cases, complex ADCs are not a good choice as they can be oversized for low speed applications, and this is the result of all the related subsystems inside of an ADC-like antialiasing filter, sampler, quantization, and encoder. The LTF in data processing quantifies light intensity variations in terms of frequency, through a current-to-frequency (CTF) converter [11,15].

An LTF generates a train of pulses with a constant duty cycle (50%) and a frequency that is a function of the irradiance incident light signal:

$$f_0 = f_D + (R_e)(E(\lambda)) \tag{6}$$

From Equation (6), it can be observed that the output frequency of the LTF f_0 is proportional to the irradiance of the perceived light $E(\lambda)$, and when no power is detected, the LTF has a constant frequency f_D, which is called dark frequency. R_e is the LTF responsivity in a certain wavelength λ and the associated units are Hz/(μW/cm^2). The irradiance is related to the surface area A_r of the LTF converter through the expression $E(\lambda) = \frac{P_{in}}{A_r}$ measured in μW/cm^2 [1,11].

The dark frequency value, f_D, results from the leak current produced by the semiconductor material and is affected by the overall system temperature [13,14].

Given that LTF output corresponds with a pulse train with variable frequency, it is important to keep in mind the different available techniques to measure it; therefore, a selection criterion of the technique will depend on the resolution and speed of the electronic interface used [14]. Thus, if a high resolution embedded system is required and time response is not too demanding, frequency counting or an accumulation of pulses can be used; if frequency is high and a high speed electronic interface is needed for measurement, given the rapid change of the light intensity, the period measurement technique is the more suitable solution [14,15]. The period measurement demands a reference clock signal with a frequency greater than the signal of interest. In the case of the TCS3200 LTF sensor, the output signal possesses frequencies between 10 Hz and 780 kHz; hence, this guides the choice of a low-cost embedded system, because almost every single chip on the market has an equipped timer with a reference signal in the order of MHz [16]. As quoted, the period measurement technique for the scenario of optical wireless communications is properly considered, given that these systems call for online processing to decrease the overall link latency.

3. Characterization of the LTF for a VLC System

In this section, we present the characterization of the LTF converter and the analysis of the proposed VLC.

In particular, an EMC 3030 HV white light LED, Tektronix TDS 3034C oscilloscope, THORLABS PM100D instrument and an LTF TCS3200 were used for the experimental setup, as shown in Figure 2. In the TCS3200, the light-to-frequency converter reads an array of 8 × 8 photodiodes with 16 photodiodes with blue filters, 16 with red filters, 16 with green filters, and the remaining 16 photodiodes are clear with no filters. For this experimental setup, the TCS3200 device was configured for its maximum output frequency and only the blue channel was used for the VLC system as the blue component of a white LED lighting has the highest bandwidth [18–20]. Given the nature of the proposed experiment, it is necessary to bear in mind that the central wavelength of the blue filter in the LTF is $\lambda_c = 470$ nm and the total area of the photodiodes is $A_r = 0.1936$ cm^2 [14].

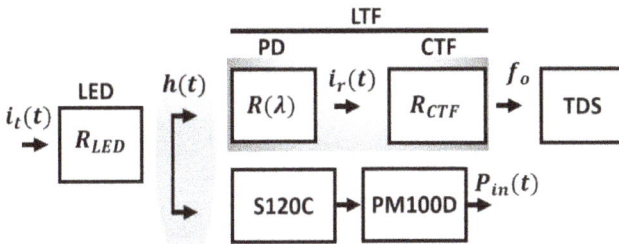

Figure 2. Experimental setup for the characterization of the LTF converter.

The schematic diagram of the experimental characterization for the LTF converter is shown in Figure 2. It consists of an optical transmitter based on a white light LED and a LTF converter acting as VLC receiver. For convenience, the LTF converter is represented as a two connected subsystems block: a photodiode and a CTF converter.

In order to analyze the performance of the LTF converter, it is necessary to derive a mathematical expression for the signal-to-noise ratio at the VLC receiver output. In this way, using Equation (1) to represent the output current of the photodiode $i_r(t)$ within the LTF block, the output signal of the CTF subsystem, f_0, can be estimated using the following expression:

$$f_0 = R_{CTF}R(\lambda)E(\lambda) + R_{CTF}n(t) \tag{7}$$

where R_{CTF} is the CTF responsivity. Now, it is important to remark that the expression for f_0, given by Equation (7), does not alter the LTF model as an analogy could be made between the terms of Equations (6) and (7), that is, $R_{CTF}R(\lambda)$ with $R_e(\lambda)$ and $R_{CTF}n(t)$ with f_D. The aforementioned statement can be demonstrated by considering an analysis of the units for each variable. The term R_{CTF} denotes a conversion factor between the input current in amperes (A) and output frequency in Hertz (Hz) of the CTF subsystem; therefore, the units of R_{CTF} are Hz/A. Next, $R(\lambda)$ is the conversion factor between the optical irradiance and the photocurrent output then its units are $\frac{A}{\mu W/cm^2}$. Thus, the term $R_{CTF}R(\lambda)$ will be given in the units of $Hz/(\mu W/cm^2)$, which is equivalent to the units of $R_e(\lambda)$ in Equation (6). Conducting a similar analysis for the term $R_{CTF}n(t)$, it can be shown that it has the same units as that of the dark frequency f_D. It is also important to note that the noise variance is scaled by the factor R_{CTF}, such that $\sigma_{f_0}^2 = R_{CTF}^2\sigma^2$, of the stochastic process $R_{CTF}n(t) = f_D$.

Once we have obtained an expression for the dark frequency f_D, the SNR value in Equation (5) can be estimated as a function of the incident irradiance and the frequency of the LTF, that is,

$$SNR = \frac{(R_{CTF}R(\lambda)E(\lambda))^2}{\sigma_{f_0}^2} = \frac{(f_0 - f_D)^2}{\sigma_{f_0}^2} \tag{8}$$

where

$$E(\lambda) = \frac{Ar}{P_{in}(t)} \tag{9}$$

3.1. Evaluation of the LTF Converter

The evaluation process for the LTF has been broken down into the following three steps:

1. constant current signal, $i_t(t)$, is applied to the transmitter LED.
2. Using the Tektronix TDS 3034C oscilloscope, the output frequency f_0 and dark frequency f_D are measured for different distance cases between the transmitter LED and the LTF.
3. Next, the incident optical power, $P_{in}(t)$, is recorded for each case using the optical sensor S120C with aperture diameter 9.5 mm, which is coupled with the THORLABS PM100D instrument. This meter console can deliver measurements of luminous flux and incident irradiance. It is not recommended to use the irradiance measurement as the PM100D instrument considers the area of the sensor S120C rather than the area of the photodiodes integrated into the LTF TCS3200 [21]. Therefore, the useful information of this experiment is the incident optical power flow $P_{in}(t)$, considering the Ar of the sensor S120C.

3.2. LTF Response to an Optical Periodic Signal

For this evaluation process, the objective is to observe the response of the LTF converter when it is excited by a periodic signal. Using this type of signal is helpful to observe the advantages and disadvantages of using the LTF as a receiver in a VLC system. For this case, the arbitrary waveform generator (AWG) RLGOL DG4162 was used to generate the modulated signal, applied to the base of the 2N3904 NPN transistor, configured in saturation mode, and acting as the driver of the LED. A frequency sweep is then carried out for the modulating signal f_{OOK} from 1 kHz until reaching the saturation frequency of the LTF. For each frequency, the separation distance of the link between the transmitter LED and the LTF was changed, from 0 cm up to the distance where the LTF output frequency was greater than or equal to the frequency of the modulating signal, that is, $f_o \geq f_{OOK}$. This limit makes sense from the viewpoint of the frequency generated by the LTF, that is, f_o reaches its maximum value during the half-period in which the modulated signal is in a high state (presence of the optical signal).

4. Results and Discussion

In this section, LTF characterization and the proposed VLC system performance analysis are evaluated, considering the input optical signal $P_{in}(t)$, variation of the link distance, LTF output frequency f_O, and SNR. We assume that the VLC channel is corrupted by AWGN. First, the LTF performance was evaluated in function of the input constant optical signal, and we proceeded with the distance variation between the transmitter LED and the LTF receiver. Figure 3 depicts LTF output frequency and optical input power versus link distance. It can be seen in Figure 3a that LTF output maximum frequency was 780 kHz (LTF saturation frequency) at the 5 mW optical input power, with 5 cm minimum link distance. On the other hand, we can see in Figure 3b that when the power input is 10 µW, the link distance that achieves the minimum output LTF frequency 1.6 kHz is 110 cm. This result is consistent with the inverse-square law, as the LED is a Lambertian source [16].

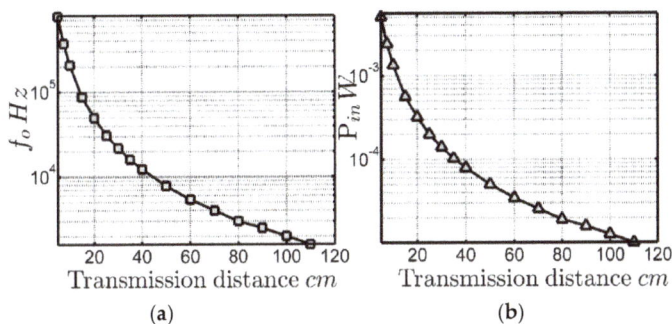

Figure 3. Experimental setup. Estimated f_O and optical power under various link distance: (**a**) LTF output frequency versus link distance; (**b**) optical input power versus link distance.

The estimated LTF responsivity value during the experiment was $R_e = 30.34$ MHz/$(\mu W/cm^2)$. This result enables the LTF to detect optical power levels of the order of nW. However, in this paper, the minimum optical power reference was limited to 10 µW, which generates a respective frequency $f_o = 2$ kHz. This configuration was important for us to experiment with a minimum frequency in the modulating signal OOK.

Additionally, based on the data presented in Figure 4, the LTF conversion factor R_e will positively affect the SNR of the system. Therefore, to generate an LTF output frequency f_o approximate to saturation, a measured SNR equal to 18.75 dB with link distance of 5 cm was found in the experiment, as illustrated in Figure 4a; for the case of less frequency $f_o = 1.6$ kHz, the SNR was around -35.15 dB, with maximum link distance of 110 cm, as illustrated in Figure 4b. The parameters estimated for the LTF are significantly different from those of the data sheet [15], because the experiment was performed under specific physical conditions and a white light LED was used.

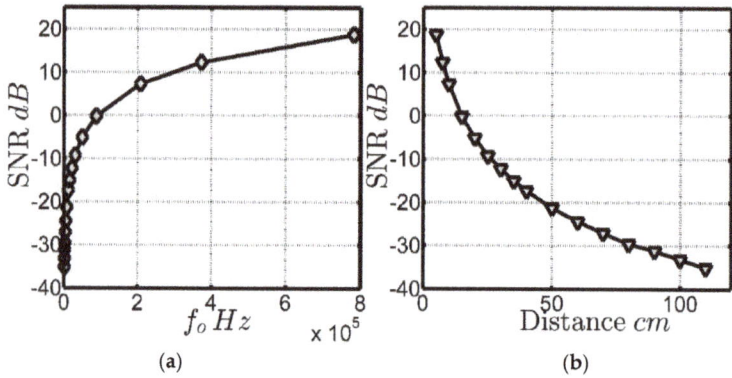

Figure 4. Model description: (a) signal-to-noise ratio (SNR) versus LTF output frequency; (b) SNR versus link distance.

The relationship $f_o = f_D$ is the dark condition (without optical power). Figure 5 summarizes the results for the dark frequency f_D versus link distance. We can see that when link distance ranges from 20 cm to 40 cm, the condition $f_D < 35$ Hz is reached, which indicates the presence of external optical sources, that is, oscilloscope, AWG, and power supplies. With this approach, it is important to mention that in the experimental setup, we do not consider focusing optical power on the LTF sensor.

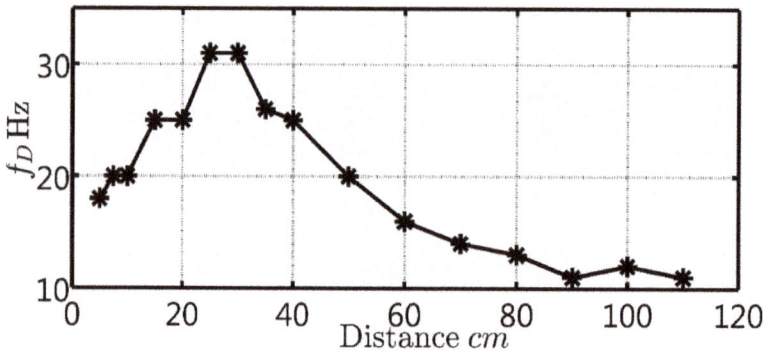

Figure 5. Dark frequency versus link distance.

The result in Figure 6 clearly shows the LTF output frequency response to light intensity variations on photodiode. At the transmitter side, the electrical OOK signal is applied to modulate the white light LED with a modulating frequency $f_{OOK} = 1$ kHz and 50% of duty cycle, as shown in Figure 6a. After free space optical transmission, the OOK signal is detected by an LTF receiver and generates an electrical signal. Then, the electrical OOK signal is converted to frequency by a current-to-frequency converter, as shown in Figure 6b.

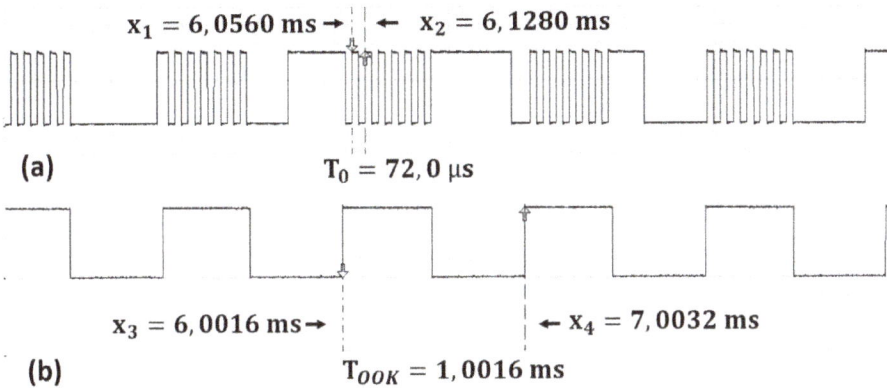

Figure 6. Transmitter with on-off keying (OOK) modulator signal and LTF as receiver signal: (**a**) optical power transmission; (**b**) LTF output frequency.

LTF converter generates a frequency around the $f_o = 13.88$ kHz when the LED transmit optical power when duty cycle is one, and, if duty cycle is zero, the LTF output frequency is $f_o = f_D$, with $f_D < 35$ Hz. The $f_D \ll f_{OOK}$; therefore, for LTF frequency estimation, it was necessary that we use the period measurement technique, for maximum data-acquisition rate (this data-acquisition rate depends on the resolution of the timer) [14,15]. However, for the VLC system, such high accuracy measurement is not necessary, because in these systems, time boundaries are wide enough to determine if a symbol is in the on-off state.

We experiment with different frequency values f_{OOK}. One thing to note, however, is the LTF frequency estimation for symbol decoding. It is necessary that the condition $f_o \geq f_{OOK}$ should be fulfilled; thus, given the unknown oscillator state of the LTF, when intensity fluctuations occur, there exists a possibility that high state of the square output signal will not be completed. Therefore, we recommend that the LTF output frequency meet the following condition $f_o \geq 4f_{OOK}$, in order to mitigate the frequency estimation problem due to the deviations generated by the LTF output.

Regarding the experiment, for each frequency f_{OOK} value, we can see the link distance between the LED and the LTF converter, which would allow finding an LTF output frequency $f_o \geq 4f_{OOK}$. Figure 7 depicts an experimental estimation of the LTF output frequency f_o versus link distance, for the different values f_{OOK}. We can see that the maximum frequency f_{OOK} of modulating signal is limited by transmission length, because the light intensity on the LTF is also a function of distance. The minimal link distance was 6.2 cm for a maximum frequency $f_{OOK} = 600$ kHz, without the LTF output frequency operates in saturation mode. Such maximum frequency could be achieved at a greater distance (>6.2 cm), if we consider an optical concentrator in the receiver. For the case of less frequency $f_{OOK} = 1$ kHz, the maximum distance was the 100 cm.

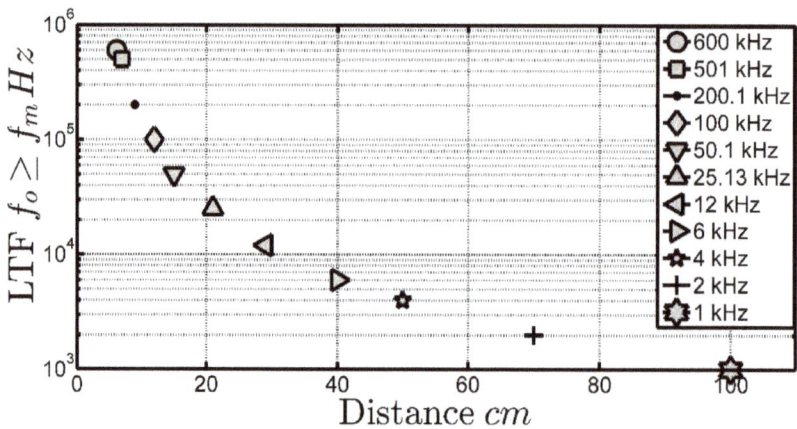

Figure 7. LTF frequency estimation for the different f_{OOK} value and variation link distance.

In Figure 4, we can appreciate the sensibility of the LTF converter to incident light. For slight variations of transmission reach, a considerate change in output frequency is observed. Nevertheless, we consider that having a receiver with high responsivity could limit the visible light communication system, if the application scenarios have light sources different to LED transmitters. Such stimulations may cause a frequency deviation of that mapped in the color-shift keying (CSK) constellation space [23].

5. Conclusions

This paper proposed a novel receiver for visible light communication system using a LTF converter as the detector. The LTF can convert the modulated light into periodic square signal with a frequency proportional to light intensity. Under this scheme, the frequency estimation can be performed with the period measurement technique, without the need of ADC modules in the receiver side. In this system, the characterization of LTF converter was presented in terms of SNR, its input optical power, and the frequency of its output squares signal. The main objective of this characterization is to identify the parameters that determine the performance of the LTF converter as a receiver in a VLC system. Among the measured parameters of the LTF detector, responsivity R_e was big enough to make the device too sensitive to light intensity variations. However, such a characteristic could affect the overall SNR of the communication, as the detector is also sensitive to power related to outside optical sources. Experimental results showed that the LTF output signal has a maximum output frequency of 600 kHz at a distance of 6.2 cm. In this setup, measured SNR reached 18.75 dB, while the lowest obtained SNR with 1.1 m length was roughly −35.1 dB. Therefore, we conclude that it is possible to use the LTF converter as a receiver in a VLC system, considering a minimum SNR value that guarantees the $f_o \geq 4f_{OOK}$ inequality. Given the condition that $f_o \geq 4f_{OOK}$, a theoretical bit rate can be achieved and would be less than or equal to 150 kbps; however, we must be aware that no bit error rate measurement was made, because the electronic elements used are noisy and under the mentioned conditions, we obtained an SNR of 18.75 dB. This could have an impact on the speed of the system, but we consider that it could be a functional VLC system, because the main application is in low data rate sceneries, so the bit rate is not critical. The $f_o \geq 4f_{OOK}$ is necessary for the identification of the LTF output frequency with the measurement system used in the receiver. Additionally, it must be ensured that the application scenario has a reduced number of non-transmitting lighting sources, in order to mitigate the negative effects of noise on the VLC receiver, and then the frequency shift in the LTF output.

The main advantage of using the LTF converter as a detector in a VLC system lies in the low complexity to convert light intensity to an electrical signal, which can be directly processed in a digital device such as a microcontroller, a field-programmer gate array (FPGA), or an embedded system equipped with a fast reference clock [14,15].

This way of detection is suggested for low data rate sceneries (indoor location, sensor networks, VLC-ID systems) because of limited bandwidth of the LTF, which is less than 800 kHz. Hence, higher data speed can be improved with an increased order of the CSK modulation format, at the cost of a penalty in transmission reach [23].

Future works include evaluation of multilevel modulation techniques like CSK [22,23] and pulse-width modulation (PWM) [24,25], which allows one to maximize the low bandwidth of the LTF converter.

Author Contributions: All authors contributed equally to writing the paper. R.A.M.C. and F.E.L.G. proposed the research project and designed the experiments; R.A.M.C. wrote the paper; A.F.B.P. contributed to the formal analysis; M.L.R. and A.F.B.P. derived the analytic model and revised the paper.

Funding: This research received no external funding.

Acknowledgments: This work was supported by the Instituto Tecnológico Metropolitano (ITM) of Colombia and Universidad Autónoma de San Luis Potosí (UASLP) of México.

Conflicts of Interest: The authors declare no conflict of interest.

References

1. Ghassemlooy, Z.; Popoola, W.; Rajbhandari, S. *Optical Wireless Communications System and Channel Modelling with MATLAB*; CRC Press: Boca Raton, FL, USA, 2013.
2. Cevik, T.; Yilmaz, S. An overview of visible light communication systems. *Int. J. Comput. Netw. Commun. (IJCNC)* **2015**, *7*, 139–150. [CrossRef]
3. Komine, T.; Nakagawa, M. Fundamental analysis for visible light communication system using LED lights. *IEEE Trans. Consum. Electron.* **2004**, *50*, 100–107. [CrossRef]
4. Grobe, L.; Paraskevopoulos, A.; Hilt, J.; Schulz, D.; Lassak, F.; Hartlieb, F.; Kottke, C.; Jungnickel, V.; Langer, K. High-speed visible light communication systems. *IEEE Commun. Mag.* **2013**, *51*, 60–66. [CrossRef]
5. Pathak, P.H.; Feng, X.T.; Hu, P.F.; Mohapatra, P. Visible light communication, networking, and sensing: A survey, potential and challenges. *IEEE Commun. Surv. Tutor.* **2015**, *17*, 2047–2077. [CrossRef]
6. Umbach, A.; Engel, T.; Bach, H.G.; van Waasen, S.; Droge, E.; Strittmatter, A.; Ebert, W.; Passenberg, W.; Steingruber, R.; Schlaak, W.; et al. Technology of InP-based 1.55-/spl mu/m ultrafast OEMMICs: 40-Gbit/s broad-band and 38/60-GHz narrow-band photoreceivers. *IEEE J. Quantum Electron.* **1999**, *35*, 1024–1031. [CrossRef]
7. IEEE Standard for Local and Metropolitan Area Networks—Part 15.7: Short-Range Wireless Optical Communication Using Visible Light, 802.15.7-2011, 2011. Available online: https://standards.ieee.org/findstds/standard/802.15.7-2011.html (accessed on 27 August 2018). [CrossRef]
8. Agrawal, G.P. *Fiber-Optic Communications Systems*; Wiley: New York, NY, USA, 2002.
9. Martinez, R.; Lopez, F.; Betancur, A. RGB sensor frequency response for a visible light communication system. *IEEE Latin Am. Trans.* **2016**, *14*, 4688–4692. [CrossRef]
10. Barrales, G.; Mocholí, S.; Vázquez, C.; Rodríguez, R.; Barrales, G. A technique for sdapting a Quasi-Digital photodetector to a frequency-to-digital converter. In Proceedings of the 2012 IEEE Ninth Electronics, Robotics and Automotive Mechanics Conference (CERMA 2012), Cuernavaca, Mexico, 19–23 November 2012; pp. 343–348.
11. Ehsan, A.; Shaari, A.; Rahman, M.; Khan, K. Optical transceiver design for POF portable optical access-card system using light-to-frequency converter. In Proceedings of the 2008 IEEE International Conference on Semiconductor Electronics, Johor Bahru, Malaysia, 25–27 November 2008; pp. 345–349.
12. Tang, F.; Shu, Z.; Ye, K.; Zhou, X.; Hu, S.; Lin, Z.; Bermak, A. A linear 126-dB dynamic range light-to-frequency converter with dark current suppression upto 125 °C for blood oxygen concentration detection. *IEEE Trans. Electron Devices* **2016**, *63*, 3983–3988. [CrossRef]
13. AMS. Light-To-Frequency—Programmable Light-To-Frequency Converter—TSL230RD. AMS Datasheet v1-00. Retrieved 14 September 2016. Available online: https://www.mouser.com/ds/2/588/TSL230RDTSL230ARDTSL230BRD-P-519226.pdf (accessed on 14 September 2016).

14. TSL230RD, TSL230ARD, TSL230BRS Programable Light-To-Frequency Converters. 2016. Available online: http://ams.com/eng/Products/Light-Sensors/Light-to-Frequency (accessed on 17 May 2016).

15. TCS3200, TCS3210, Programmable Color Light-To-Frequency Converter. 2009. Available online: http://ams.com/eng/Products/Light-Sensors/Color-Sensors/TCS3200 (accessed on 10 July 2015).

16. Yurish, S. Intelligent opto sensors' interfacing based on universal. frequency-to-digital converter. *Sens. Transducers Mag.* **2005**, *56*, 326–334.

17. Monteiro, E. Design and Implementation of Color-Shift Keying for Visible Light Communications. Master's Thesis, McMaster University, Hamilton, ON, Canada, 2013.

18. Bhalerao, M.; Sonavane, S.; Kumar, V. A survey of wireless communication using visible light. *Int. J. Adv. Eng. Technol.* **2013**, *5*, 188–197.

19. Li, H.; Chen, X.; Huang, B.; Tang, D.; Chen, H. High bandwidth visible light communications based on a post-equalization circuit. *IEEE Photonics Technol. Lett.* **2014**, *26*, 119–122. [CrossRef]

20. Zeng, L.; O'Brien, D.; Le-Minh, H.; Lee, K.; Jung, D.; Oh, Y. Improvement of date rate by using equalization in an indoor visible light communication system. In Proceedings of the 2008 4th IEEE International Conference on Circuits and Systems for Communications, Shanghai, China, 26–28 May 2008; pp. 678–682.

21. THORLABS. PM100D Compact Power and Energy Meter Console. 2016. Available online: https://www.thorlabs.com/thorproduct.cfm?partnumber=PM100D (accessed on 28 September 2016).

22. Murata, N.; Kozawa, Y.; Umeda, Y. Digital color shift keying with multicolor LED array. *IEEE Photonics J.* **2016**, *8*, 1–13. [CrossRef]

23. Luna-Rivera, J.; Suarez-Rodriguez, C.; Guerra, V.; Perez-Jimenez, R.; Rabadan-Borges, J.; Rufo-Torres, J. Low-complexity colour-shift keying-based visible light communications system. *IET Optoelectron.* **2015**, *9*, 191–198. [CrossRef]

24. Assabir, A.; Elmhamdi, J.; Hammouch, A.; Akherraz, A. Application of Li-Fi technology in the transmission of the sound at the base of the PWM. In Proceedings of the 2016 International Conference on Electrical and Information Technologies (ICEIT), Tangiers, Morocco, 4–7 May 2016; pp. 260–265.

25. Pradana, A.; Ahmadi, N.; Adionos, T. Design and implementation of visible light communication system using pulse width modulation. In Proceedings of the 2015 International Conference on Electrical Engineering and Informatics (ICEEI), Denpasar, Indonesia, 10–11 August 2015; pp. 25–30.

electronics

MDPI

Article

Adaptive Power Allocation Scheme for Mobile NOMA Visible Light Communication System

Zanyang Dong *, Tao Shang *, Qian Li and Tang Tang

State Key Laboratory of Integrated Service Networks, School of Telecommunications Engineering,
Xidian University, Xi'an 710071, China; liqian_xd@foxmail.com (Q.L.); tangt_xd@foxmail.com (T.T.)
* Correspondence: zanydong@foxmail.com (Z.D.); tshang@xidian.edu.cn (T.S.);
   Tel.: +86-187-1084-6102 (Z.D.); +86-134-6893-5375 (T.S.)

Received: 27 February 2019; Accepted: 26 March 2019; Published: 29 March 2019

Abstract: Recently, due to its higher spectral efficiency and enhanced user experience, non-orthogonal multiple access (NOMA) has been widely studied in visible light communication (VLC) systems. As a main concern in NOMA-VLC systems, the power allocation scheme greatly affects the tradeoff between the total achievable data rate and user fairness. In this context, our main aim in this work was to find a more balanced power allocation scheme. To this end, an adaptive power allocation scheme based on multi-attribute decision making (MADM), which flexibly chooses between conventional power allocation or inverse power allocation (IPA) and the optimal power allocation factor, has been proposed. The concept of IPA is put forward for the first time and proves to be beneficial to achieving a higher total achievable data rate at the cost of user fairness. Moreover, considering users' mobility along certain trajectories, we derived a fitting model of the optimal power allocation factor. The feasibility of the proposed adaptive scheme was verified through simulation and the fitting model was approximated to be the sum of three Gaussian functions.

Keywords: visible light communication; non-orthogonal multiple access; inverse power allocation scheme; adaptive power allocation scheme; fitting model

---

1. Introduction

Due to the ever-increasing penetration of wireless communication devices such as smartphones and tablets, rapidly growing wireless data traffic is expected to exceed 500 exabytes by 2020 [1], which is placing pressure on the dwindling radio frequency (RF) spectrum. Along with considerable advances in solid-state lighting, visible light communication (VLC) [2–4] supporting remarkably high-speed wireless communication has attracted great attention as a promising technology in applications such as the Internet of Things (IoT), 5G systems, underwater communications, vehicle-to-vehicle communication, and so on. In addition to the nature of its wide available bandwidth, VLC also features low power consumption, an unlicensed spectrum, enhanced confidentiality, and anti-electromagnetic interference, etc.

In VLC, it is essential to ensure the core functionality of providing multiple users with ubiquitous connectivity as well as broadband communication. To this effect, an appropriate multiple access (MA) scheme should be involved in dealing with simultaneous network access requests from multiple users. Traditionally, orthogonal multiple access (OMA) schemes have been applied to VLC systems, including carrier sense multiple access (CSMA), code division multiple access (CDMA), and orthogonal frequency division multiple access (OFDMA) [5]. Recently, a spectrum-efficient multiple access scheme called non-orthogonal multiple access (NOMA) has been proposed to further enhance system capacity and achieve a better balance between system fairness and throughput [6,7]. As a promising solution for next generation wireless networks, NOMA allocates different power levels to each user based on its channel condition, thus achieving power-domain multiplexing of multiple users. Differently from a traditional OMA system, NOMA allow users to share all time-frequency (TF) resources and has proven to be superior theoretically and experimentally [8]. Apart from its applications in RF communications, NOMA has been introduced to VLC systems [9] and abundant research achievements have been obtained [10–13], especially those concerning power allocation schemes. In the literature by L. Yin et al. [14], the performance of NOMA-VLC was investigated based on a fixed power allocation (FPA) scheme. In addition, a channel-dependent gain ratio power allocation (GRPA) scheme was proposed in [9], which turned out to be superior to the FPA scheme. In addition, two types of quality of service (QoS) guaranteed power allocation have been proposed to iteratively optimize the sum user rate or max-min user rate utilizing gradient projection (GP) algorithm [15]. However, in all the existing works, users with a lower channel gain are always allocated a higher power level, which has been regarded as a basic principle in NOMA and has been proven to be beneficial to user fairness.

In our work, we begin with a hypothesis, which can be called inverse power allocation (IPA), that the total achievable data rate may be higher if users with a worse channel condition are allocated less power. We then prove this through theoretical formulas and simulation analysis. However, the total achievable data rate gain was obtained at a cost of user fairness. Hence, in order to achieve a better balance between total achievable data rate and user fairness, we attempted to find an adaptive power allocation scheme with which to combine conventional power allocation and IPA flexibly. To this effect, a multi-attribute decision making (MADM) algorithm was adopted to choose a suitable scheme, i.e., conventional or IPA, and an optimal power allocation factor in real time according to a mathematical comprehensive assessment of the total achievable data rate and user fairness. Moreover, by assuming users walk through certain trajectories within the optical attocell, we obtained a fitting model of the optimal power allocation factor utilizing the curve fitting technique.

The contribution of this paper is three-fold: first, to the best of our knowledge, this is the first work involving IPA in NOMA-VLC systems; second, an adaptive power allocation scheme based on MADM is proposed, which effectively combines IPA with conventional power allocation and facilitates the choice of an optimal power allocation factor; and, finally, taking users' mobility into account, a fitting model of optimal power allocation factor is presented.

The remainder of the paper is organized as follows: Section 2 illustates the model of the NOMA-VLC system. In Section 3, the IPA scheme is presented and the effect of it on system performance is discussed. An overall adaptive power allocation scheme is proposed in Section 4. In Section 5, the simulation results and discussion are presented. The modeling of the optimal power allocation factor for a mobile NOMA-VLC system is presented in Section 6. Finally, Section 7 concludes the paper.

2. System Model

Figure 1 shows the NOMA-VLC system model. All the devices were purchased from Vishay (Tianjin, China). For illustrative purposes, we consider a single optical attocell deployment and mainly focus on the NOMA downlink in an indoor environment, in which one light-emitting-diode (LED) transmitter is installed and M users are served. The LED transmitter can not only provide illumination but also convert electrical signals, which derive from a power line communications (PLC) backbone network, into optical signals by modulating the intensity of the emitted light. In addition, each user is equipped with a single photodiode (PD), which is used for extraction of the transmitted signal from the received optical carrier by direct detection. As is shown in Figure 1, R denotes the maximum cell radius, H denotes the vertical distance from the LED to the receiving plane of the users, and r_k denotes the horizontal separation from the k-th user to the LED.

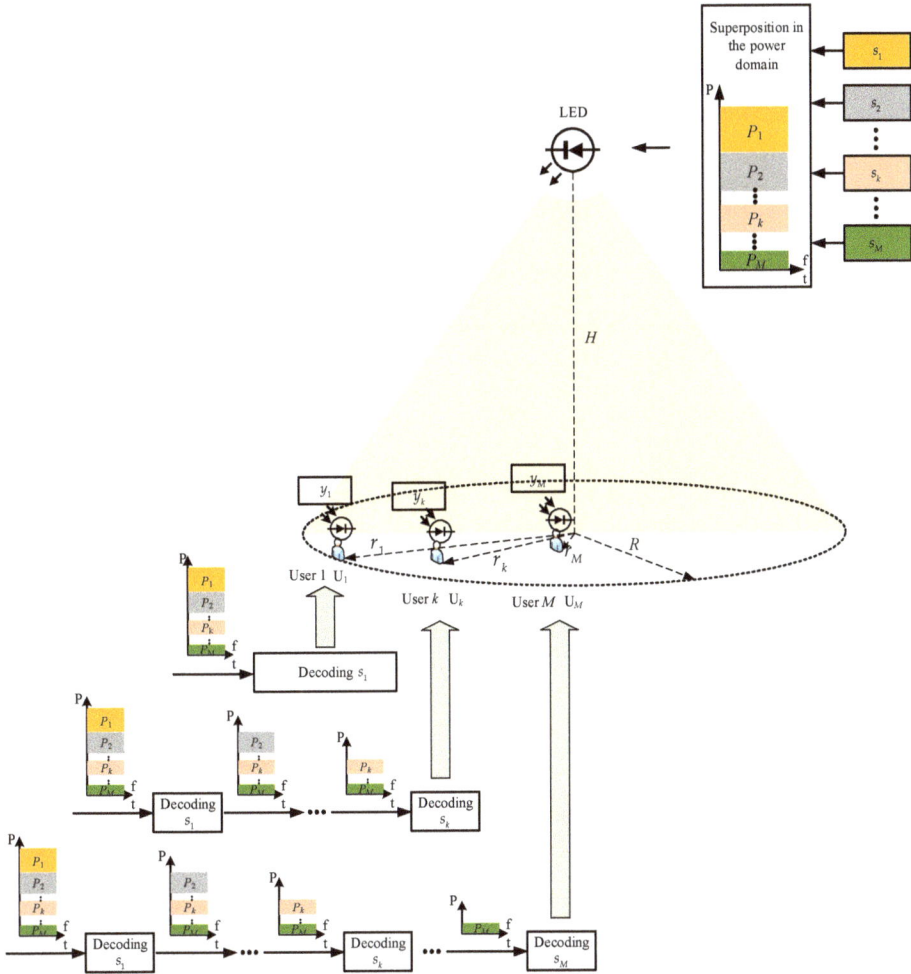

Figure 1. System model of non-orthogonal multiple access visible light communication (NOMA-VLC). Legend: LED, light-emitting-diode.

We assume that the LED follows a generalized Lambertian radiation pattern and the PD at each user faces vertically upwards with the width of the field of view denoted by ψ_{FOV}. Due to the weakness of diffuse components, which have proven to be at least 7 dB lower than the line of sight (LOS) component [16], the direct current (DC) channel gain for the k-th user can be approximately calculated by considering the LOS link, the wideband nature of VLC, and the shadowing effect:

$$
\begin{aligned}
h_k &= \frac{(m+1)AR_p}{2\pi d_k^2} \cos^m(\phi_k) \cos(\psi_k) T_s(\psi_k) g(\psi_k) \\
&= \frac{AR_p(m+1)H^{m+1}T_s(\psi_k)g(\psi_k)}{2\pi(r_k^2+H^2)^{m+3/2}}
\end{aligned}
\tag{1}
$$

Here, m denotes the order of the Lambertian radiation pattern, which is derived from the semi-angle of the LED, $\Phi_{1/2}$, as $m = -1/\log_2(\cos(\Phi_{1/2}))$; A denotes the physical area of the PD; R_p denotes the responsivity of the PD; d_k denotes the Euclidean distance between the k-th user and the LED; ϕ_k denotes the angle of irradiance at the k-th user; ψ_k denotes the angle of incidence at the k-th user; $T_s(\psi_k)$ denotes the gain of the optical filter used at the receiver; and $g(\psi_k)$ denotes the gain of the optical concentrator used at the receiver front-end, which is given by [17]

$$g(\psi_k) = \begin{cases} \frac{n^2}{\sin^2 \psi_{FOV}} & 0 \leq \varphi_k \leq \psi_{FOV} \\ 0 & \varphi_k \geq \psi_{FOV}, \end{cases} \tag{2}$$

where n denotes the refractive index of the optical concentrator.

Without loss of generality, based on the DC channel gains, all users $U_1, \cdots, U_M$ can be sorted in an ascending order as $h_1 \leq \cdots \leq h_k \leq \cdots \leq h_M$.

The principle of NOMA is also illustrated in Figure 1. At the transmitter side, the messages $\{s_i, i = 1, 2, \cdots M\}$ intended for all the corresponding users are superposed in the power domain with associated power values $\{P_i, i = 1, 2, \cdots M\}$ and then transmitted simultaneously as

$$x = \sum_{i=1}^{M} a_i \sqrt{P_{elec}} s_i + I_{DC}, \tag{3}$$

where x denotes the superposed signal of $\{s_i, i = 1, 2, \cdots M\}$; P_{elec} denotes the total electrical power of all the message signals; I_{DC} denotes a DC bias added before transmission to ensure the positive signal; and a_i denotes the power allocation coefficient for the i-th user, which should satisfy the following two conditions: $\sum_{i=1}^{M} a_i^2 = 1$, according to the total power constraint; and $a_1 \geq \cdots \geq a_k \geq \cdots \geq a_M$, according to the basic principle of conventional NOMA.

Similarly to the power allocation coefficient, another equivalent parameter, the power allocation factor α, can be defined as $\alpha = a_i^2 / a_{i-1}^2, i = 2, \cdots M$, which can be constant or variable along with i according to different power allocation schemes, i.e., FPA or GRPA.

At the receiver side of the k-th user, taking the DC channel gain and the additive white Gaussian noise (AWGN) into account and removing the DC term, we can obtain the received signal

$$y_k = \sqrt{P_{elec}} h_k \left(\sum_{i=1}^{M} a_i s_i \right) + z_k, \tag{4}$$

where y_k denotes the received signal at the k-th user and z_k denotes the AWGN with a zero mean and variance σ_k^2. Moreover, $\sigma_k^2 = N_0 B$, where N_0 denotes the noise power spectral density and B denotes the channel bandwidth. Next, the successive interference cancellation (SIC) is performed to extract s_k from the received signal, with the process being as follows: first, we attempt to obtain the message signal s_1 intended for the first user, with the other signals treated as noise; then, by subtracting s_1 from the received signal, with the residual interference fraction denoted by ε [18], and treating the message signal for the users with stronger channel gains than the second user as noise, we can obtain the message signal s_2; and finally, by following the former method, $s_3, \cdots, s_{k-1}, s_k$ are obtained in sequence. According to the Shannon Theorem, the achievable data rate for the k-th user may be given by

$$R_k = \begin{cases} \frac{B}{2} \log_2 \left(1 + \frac{(h_k a_k)^2}{\sum_{i=1}^{k-1} \varepsilon (h_k a_i)^2 + \sum_{j=k+1}^{M} (h_k a_j)^2 + 1/\rho} \right) & k = 1, \cdots, M-1 \\ \frac{B}{2} \log_2 \left(1 + \frac{(h_k a_k)^2}{\sum_{i=1}^{k-1} \varepsilon (h_k a_i)^2 + 1/\rho} \right) & k = M \end{cases} \tag{5}$$

where $\rho = P_{elec} / (N_0 B)$ and the scaling factor $1/2$ comes from the constraint of the real-valued signal, i.e., Hermitian symmetry.

3. Inverse Power Allocation Scheme

We define the features of an IPA scheme as follows: first, at the transmitter side, users with a worse channel condition are allocated less power, and second, at the receiver side, the message signal intended for users with a worse channel condition has a higher decoding order. The differences and links between the IPA scheme and the conventional power allocation scheme are illustrated in Figure 2, in which we assume that there are two users for simplicity, i.e., $M = 2$.

Let a_i, a_i denote the power allocation coefficient for user i in the conventional power allocation case and the IPA case, respectively, where $i = 1, 2$. As mentioned above, $a_1 \geq a_2$, $a_1^2 + a_2^2 = 1$, $a_1' \leq a_2'$, and $a_1'^2 + a_2'^2 = 1$. In addition, the power allocation factor α, which should be less than 1, can be described as a_2^2/a_1^2 or $a_1'^2/a_2'^2$ in the conventional power allocation case or the IPA case, respectively. Moreover, Equations (6) and (7) can be easily derived for the below two cases.

$$\begin{cases} a_1^2 = \frac{1}{1+\alpha} \\ a_2^2 = \frac{\alpha}{1+\alpha}, \end{cases} \tag{6}$$

$$\begin{cases} a_1'^2 = \frac{\alpha}{1+\alpha} \\ a_2'^2 = \frac{1}{1+\alpha}. \end{cases} \tag{7}$$

According to Equations (5) and (6), the total achievable data rate of the two users in the conventional power allocation case can be given by:

$$\begin{aligned} R_{total} &= \frac{B}{2} \log_2(1 + \frac{(h_1 a_1)^2}{(h_1 a_2)^2 + 1/\rho}) + \frac{B}{2} \log_2(1 + \frac{(h_2 a_2)^2}{\varepsilon(h_2 a_1)^2 + 1/\rho}) \\ &= \frac{B}{2} \log_2[(1 + \frac{(h_1 a_1)^2}{(h_1 a_2)^2 + 1/\rho})(1 + \frac{(h_2 a_2)^2}{\varepsilon(h_2 a_1)^2 + 1/\rho})] \\ &= \frac{B}{2} \log_2[(1 + \frac{h_1^2}{h_1^2 \alpha + (\alpha+1)/\rho})(1 + \frac{h_2^2 \alpha}{\varepsilon h_2^2 + (\alpha+1)/\rho})] \end{aligned} \tag{8}$$

Similarly, the total achievable data rate of the two users in the IPA case can be given by:

$$\begin{aligned} R'_{total} &= \frac{B}{2} \log_2(1 + \frac{(h_1 a_1')^2}{\varepsilon(h_1 a_2')^2 + 1/\rho}) + \frac{B}{2} \log_2(1 + \frac{(h_2 a_2')^2}{(h_2 a_1')^2 + 1/\rho}) \\ &= \frac{B}{2} \log_2[(1 + \frac{(h_1 a_1')^2}{\varepsilon(h_1 a_2')^2 + 1/\rho})(1 + \frac{(h_2 a_2')^2}{(h_2 a_1')^2 + 1/\rho})] \\ &= \frac{B}{2} \log_2[(1 + \frac{h_1^2 \alpha}{\varepsilon h_1^2 + (\alpha+1)/\rho})(1 + \frac{h_2^2}{h_2^2 \alpha + (\alpha+1)/\rho})] \end{aligned} \tag{9}$$

Next, we carry out a numerical simulation utilizing MATLAB R2016a to intuitively show the size relationship of the total achievable data rate in these two cases; the simulation setup is parameterized as Table 1. In this setup, the PD parameters are set according to BPW21R [19], a planar Silicon PN photodiode in a hermetically sealed short TO-5 case. In addition, the simulation step of the horizontal separation from each user to the LED, i.e., r_1 and r_2, is set to 0.1 m. In accordance with [9], we chose $\alpha = 0.3$ and $\alpha = 0.4$, which have been proven to be optimal to achieving the best performance when an FPA scheme is adopted. Figure 3 shows the simulation results.

As is shown in Figure 3, the IPA scheme leads to a higher total achievable data rate compared with the conventional power allocation scheme whether $\alpha = 0.3$ or $\alpha = 0.4$, and the gain increases with a decrease in α. However, it is obvious that the IPA scheme will cause serious unfairness between the two users. Hence, adopting the IPA scheme solely for the NOMA-VLC system is not optimal; this encouraged us to propose an adaptive power allocation scheme to achieve a better balance between total achievable data rate and user fairness by choosing a suitable scheme, i.e., conventional or inverse, and an optimal power allocation factor.

Conventional power allocation scheme

(a)

IPA scheme

(b)

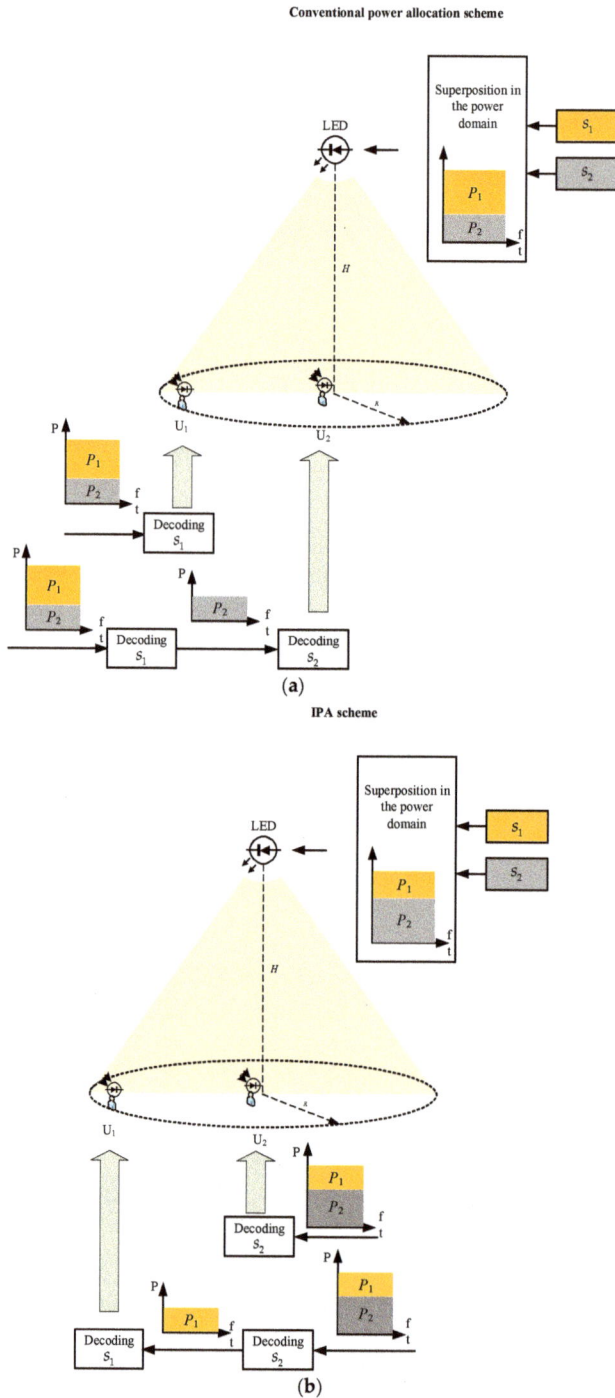

Figure 2. NOMA-VLC system based on different power allocation schemes: (**a**) conventional power allocation scheme; (**b**) inverse power allocation (IPA) scheme.

Table 1. Simulation parameters. Legend: PD, photodiode.

Parameter Name, Notation	Value
Vertical height, H	3 m
LED semi-angle, $\Phi_{1/2}$	60°
Signal power, P_{elec}	1.25 mW
Channel bandwidth, B	20 MHz
Noise power spectral density, N_0	10^{-21} A^2/Hz
PD physical area, A	7.5 mm^2
PD responsivity, R_p	0.48 A/W
PD's field of view (FOV), ψ_{FOV}	50°
Optical filter gain, $T_s(\psi_k)$	1
Refractive index, n	1.5
Power allocation factor, α	0.3
Residual interference fraction, ε	0.1

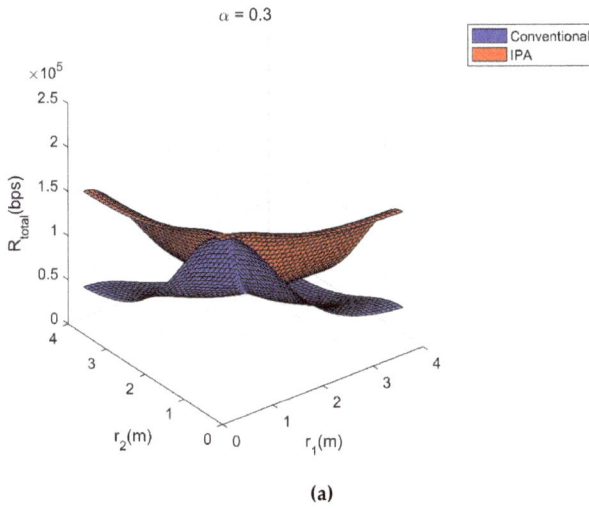

(a)

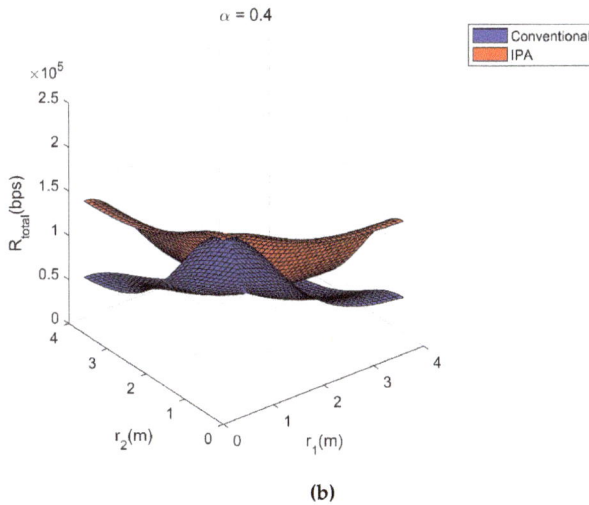

(b)

Figure 3. Total achievable data rate in conventional power allocation case and IPA case when: (a) $\alpha = 0.3$; (b) $\alpha = 0.4$.

4. Adaptive Power Allocation Scheme

In this section, we propose an adaptive power allocation scheme based on MADM, wherein the total achievable data rate and user fairness are selected as decision parameters. Our goal was to choose the most appropriate combination of a power allocation scheme and corresponding power allocation factor, and the change in users' location. Moreover, the choice space is a $u \times v$ matrix, where u is the number of candidate schemes, i.e., $u = 2$, and v is the number of candidate power allocation factors, which are discretized artificially. The concrete implementation process is as follows: first, the standard deviation method [20], which uses mathematical variance information to solve the MADM problem, is used to obtain the objective weight of each decision parameter; then, the technique for order preference by similarity to the ideal solution (TOPSIS) [21], which makes the most of the information of the raw data, is used to sort the candidate combinations in order to choose the best one.

According to the standard deviation method, the main steps used to obtain the objective weight of each decision parameter are as follows:

First, the normalized decision matrix C is constructed as

$$
C = \begin{pmatrix} C_{11} & C_{12} \\ C_{21} & C_{22} \\ \cdots & \cdots \\ \cdots & \cdots \\ C_{p1} & C_{p2} \end{pmatrix}, \tag{10}
$$

where p is the total number of candidate combinations and $p = u \times v = 2v$, and elements C_{k1}, C_{k2} are the normalized values of total achievable data rate and user fairness, respectively, when the k-th candidate combination is chosen. As for the benefit parameters, their normalization can be given by [22]

$$
C_{ki} = \frac{S_{ki} - \min(S_{xi}, 1 \leq x \leq p)}{\max(S_{xi}, 1 \leq x \leq p) - \min(S_{xi}, 1 \leq x \leq p)}, \quad 1 \leq k \leq p, i = 1, 2, \tag{11}
$$

where S_{k1} is the value of the total achievable data rate when the k-th candidate combination is chosen, which can be calculated as Equations (8) and (9), and S_{k2} is the value of user fairness when the k-th candidate combination is chosen, which can be given by

$$
S_{k2} = \frac{\min(R_1|k, R_2|k)}{\max(R_1|k, R_2|k)}
$$

$$
= \begin{cases} \dfrac{\min(\frac{B}{2}\log_2(1+\frac{h_1{}^2}{h_1{}^2\alpha+(\alpha+1)/\rho}), \frac{B}{2}\log_2(1+\frac{h_2{}^2\alpha}{eh_2{}^2+(\alpha+1)/\rho}))}{\max(\frac{B}{2}\log_2(1+\frac{h_1{}^2}{h_1{}^2\alpha+(\alpha+1)/\rho}), \frac{B}{2}\log_2(1+\frac{h_2{}^2\alpha}{eh_2{}^2+(\alpha+1)/\rho}))}, \\ \quad k \leq p/2, r_1 \geq r_2, \text{or } k > p/2, r_1 \leq r_2 \\[2ex] \dfrac{\min(\frac{B}{2}\log_2(1+\frac{h_1{}^2\alpha}{eh_1{}^2+(\alpha+1)/\rho}), \frac{B}{2}\log_2(1+\frac{h_2{}^2}{h_2{}^2\alpha+(\alpha+1)/\rho}))}{\max(\frac{B}{2}\log_2(1+\frac{h_1{}^2\alpha}{eh_1{}^2+(\alpha+1)/\rho}), \frac{B}{2}\log_2(1+\frac{h_2{}^2}{h_2{}^2\alpha+(\alpha+1)/\rho}))}, \\ \quad k \leq p/2, r_1 \leq r_2, \text{or } k > p/2, r_1 \geq r_2, \end{cases} \tag{12}
$$

where $R_1|k, R_2|k$ is the achievable data rate of User 1 and User 2, respectively, when the k-th candidate combination is chosen.

Second, the objective weight of each decision parameter is calculated as

$$
w_j = \frac{\sqrt{\sum\limits_{i=1}^{p}\left(C_{ij} - \frac{1}{p}\sum\limits_{i=1}^{p}C_{ij}\right)^2 / (p-1)}}{\sum\limits_{j=1}^{2}\sqrt{\sum\limits_{i=1}^{p}\left(C_{ij} - \frac{1}{p}\sum\limits_{i=1}^{p}C_{ij}\right)^2 / (p-1)}}, \quad j = 1, 2, \tag{13}
$$

where w_1, w_2, are the objective weights of the total achievable data rate and user fairness, respectively.

Once we obtain the objective weight of each parameter, according to TOPSIS, the main steps which must be used to choose the best candidate combination are as follows:

1. Construct the weighted normalized decision matrix D as:

$$D = \begin{pmatrix} D_{11} & D_{12} \\ D_{21} & D_{22} \\ \cdots & \cdots \\ \cdots & \cdots \\ D_{p1} & D_{p2} \end{pmatrix} = \begin{pmatrix} w_1 C_{11} & w_2 C_{12} \\ w_1 C_{21} & w_2 C_{22} \\ \cdots & \cdots \\ \cdots & \cdots \\ w_1 C_{p1} & w_2 C_{p2} \end{pmatrix} \quad (14)$$

2. Determine the positive ideal solution matrix Y^+ as:

$$Y^+ = \begin{pmatrix} Y_1^+ & Y_2^+ \end{pmatrix} = \begin{pmatrix} \max_k(D_{k1}) & \max_k(D_{k2}) \end{pmatrix}, \quad k = 1, 2, \cdots, p. \quad (15)$$

3. Determine the negative ideal solution matrix Y^- as:

$$Y^- = \begin{pmatrix} Y_1^- & Y_2^- \end{pmatrix} = \begin{pmatrix} \min_k(D_{k1}) & \min_k(D_{k2}) \end{pmatrix}, \quad k = 1, 2, \cdots, p. \quad (16)$$

4. Calculate the Euclidean distance between each solution and the positive ideal solution as:

$$F_k^+ = \sqrt{\sum_{i=1}^{2} (D_{ki} - Y_i^+)^2}, \quad k = 1, 2, \cdots, p. \quad (17)$$

5. Calculate the Euclidean distance between each solution and the negative ideal solution as:

$$F_k^- = \sqrt{\sum_{i=1}^{2} (D_{ki} - Y_i^-)^2}, \quad k = 1, 2, \cdots, p. \quad (18)$$

6. Calculate the relative proximity of each solution to the ideal solution as:

$$G_k = \frac{F_k^-}{F_k^+ + F_k^-}, \quad 0 \le G_k \le 1, k = 1, 2, \cdots, p. \quad (19)$$

7. Find the best combination of a power allocation scheme and the corresponding power allocation factor by:

$$\operatorname*{argmax}_k G_k, \quad k = 1, 2, \cdots, p. \quad (20)$$

Next, we extend the adaptive power allocation scheme to adapt to more realistic scenarios in which M users exist and $M > 2$. For ease of identification, we further define $\alpha_{(i-1)i}$ to describe the power allocation factor between the $(i-1)$-th user and the i-th one. As mentioned before, $\alpha_{(i-1)i}$ can be expressed as $\alpha_{(i-1)i} = a_i^2 / a_{i-1}^2, i = 2, \cdots M$. The concrete process with which to obtain the optimal $\alpha_{(i-1)i}, i = 2, \cdots M$ is presented below.

First, we use α_{12}, which is now a variable to be optimized, to express all the other power allocation factors, namely, $\alpha_{23}, \cdots, \alpha_{(M-1)M}$. According to GRPA, $\alpha_{(i-1)i} = (h_1/h_i)^i, i = 2, \cdots, M$. Based on this equation, it is easy to obtain the recursion relation of the power allocation factor: $\alpha_{i(i+1)}/\alpha_{(i-1)i} = h_1 h_i^i / h_{i+1}^{i+1}, i = 2, \cdots, M$. Following this recursion relation, we can easily express $\alpha_{23}, \cdots, \alpha_{(M-1)M}$ in terms of α_{12}. Next, we obtain the optimal α_{12} by means of the proposed adaptive scheme, in which the Equations (8), (9) and (12) need to be extended to take the effect of all users on the decision parameters, namely, the total achievable data rate and user fairness, into account. Specifically, Equations (8), (9) and (12) are extended as Equations (21)–(23), respectively.

$$
\begin{aligned}
R_{\text{total}} &= \sum_{k=1}^{M} R_k \\
&= \frac{B}{2} \log_2 \left(1 + \frac{(h_M a_M)^2}{\sum\limits_{i=1}^{M-1} \epsilon(h_M a_i)^2 + 1/\rho}\right) + \sum_{k=1}^{M-1} \frac{B}{2} \log_2 \left(1 + \frac{(h_k a_k)^2}{\sum\limits_{i=1}^{k-1} \epsilon(h_k a_i)^2 + \sum\limits_{j=k+1}^{M} (h_k a_j)^2 + 1/\rho}\right),
\end{aligned}
\tag{21}
$$

$$
\begin{aligned}
R'_{\text{total}} &= \sum_{k=1}^{M} R'_k \\
&= \frac{B}{2} \log_2 \left(1 + \frac{(h_1 a'_1)^2}{\sum\limits_{i=2}^{M} \epsilon(h_1 a'_i)^2 + 1/\rho}\right) + \sum_{k=2}^{M} \frac{B}{2} \log_2 \left(1 + \frac{(h_k a'_k)^2}{\sum\limits_{i=k+1}^{M} \epsilon(h_k a'_i)^2 + \sum\limits_{j=1}^{k-1} (h_k a'_j)^2 + 1/\rho}\right),
\end{aligned}
\tag{22}
$$

$$
S_{k2} = \frac{\min(R_1|k, R_2|k, \cdots, R_M|k)}{\max(R_1|k, R_2|k \cdots, R_M|k)} \; or \; \frac{\min(R'_1|k, R'_2|k, \cdots, R'_M|k)}{\max(R'_1|k, R'_2|k \cdots, R'_M|k)}, Conventional \; or \; IPA,
\tag{23}
$$

where a_k, a'_k can be expressed as follows:

$$
a_k = \begin{cases}
\sqrt{1/(1 + \alpha_{12} + \alpha_{12} \times \alpha_{23} + \cdots + \alpha_{12} \times \alpha_{23} \times \cdots \times \alpha_{(M-1)M})} & , k = 1 \\
\sqrt{\alpha_{12} \times \alpha_{23} \times \cdots \times \alpha_{(k-1)k}/(1 + \alpha_{12} + \alpha_{12} \times \alpha_{23} + \cdots + \alpha_{12} \times \alpha_{23} \times \cdots \times \alpha_{(M-1)M})} & , k = 2, \cdots, M
\end{cases}
\tag{24}
$$

$$
a'_k = \begin{cases}
\sqrt{1/(1 + \alpha_{(M-1)M} + \alpha_{(M-1)M} \times \alpha_{(M-2)(M-1)} + \cdots + \alpha_{12} \times \alpha_{23} \times \cdots \times \alpha_{(M-1)M})} \\
\quad , k = M \\
\sqrt{\alpha_{k(k+1)} \times \cdots \times \alpha_{(M-1)M}/(1 + \alpha_{(M-1)M} + \alpha_{(M-1)M} \times \alpha_{(M-2)(M-1)} + \cdots + \alpha_{12} \times \alpha_{23} \times \cdots \times \alpha_{(M-1)M})} \\
\quad , k = 1, \cdots, M - 1.
\end{cases}
\tag{25}
$$

Substituting the expressions of $\alpha_{23}, \cdots, \alpha_{(M-1)M}$ into (24) and (25), we can also obtain a_k, a'_k in terms of α_{12}. Finally, once the optimal α_{12} is obtained, we can easily determine the optimal $\alpha_{23}, \cdots, \alpha_{(M-1)M}$ in turn.

In order to make the proposed adaptive power allocation scheme clearer, a flow chart of the specific implementation process is given in Figure 4.

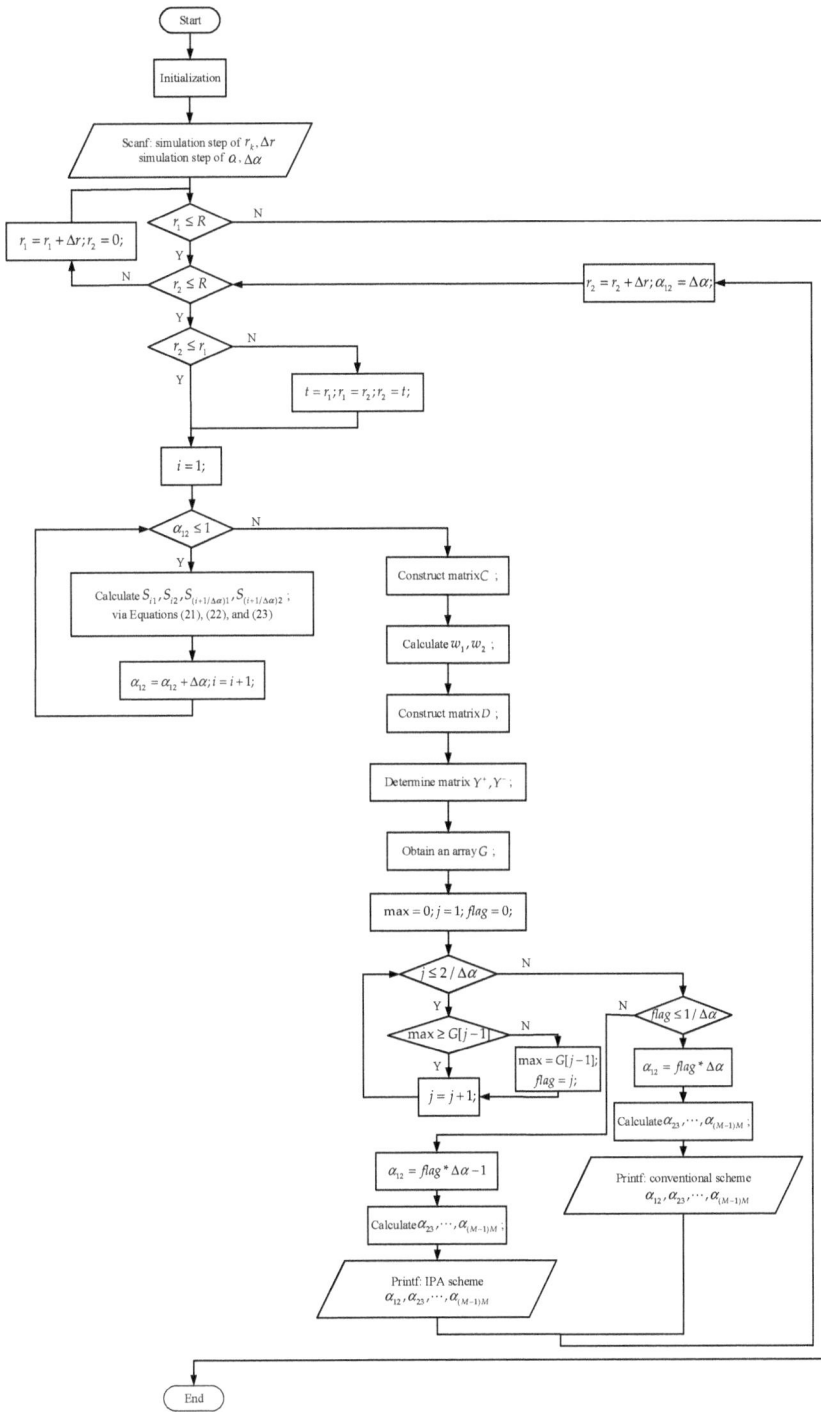

Figure 4. Flow chart of the proposed adaptive power allocation scheme.

5. Simulation Results and Discussion

In order to verify the feasibility of our proposed adaptive power allocation scheme, we conducted a simulation analysis utilizing matrix laboratory (MATLAB) R2016a, and three scenarios were chosen as examples: Scenario 1, $M = 2$; Scenario 2, $M = 5$; and Scenario 3, $M = 10$. In addition, the simulation step r_k was still set to 0.1 m and the simulation step of α was set to 0.01. Other parameters were the same as in Table 1. Moreover, we also simulated these three scenarios with a GRPA scheme and IPA scheme, respectively, in order to verify the superiority of our proposed scheme compared with them. For Scenario 1, the simulation results are shown in Figure 5 with the ergodic positions of both users taken into account. For Scenario 2 and Scenario 3, we randomly chose the combination of positions of all users involved and tested each scheme ten times, with the simulation results shown in Figures 6 and 7.

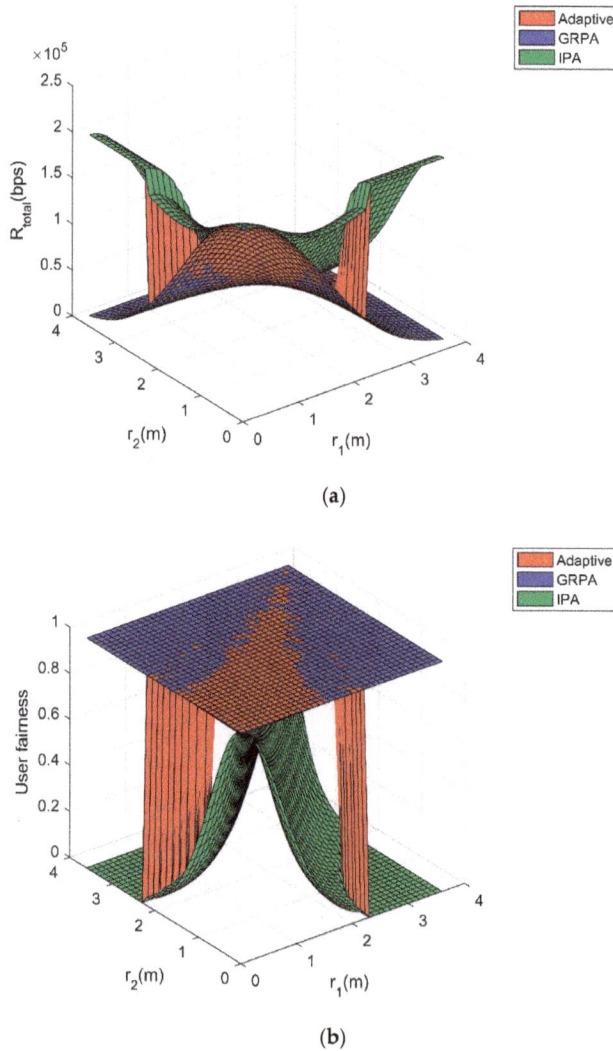

(a)

(b)

Figure 5. *Cont.*

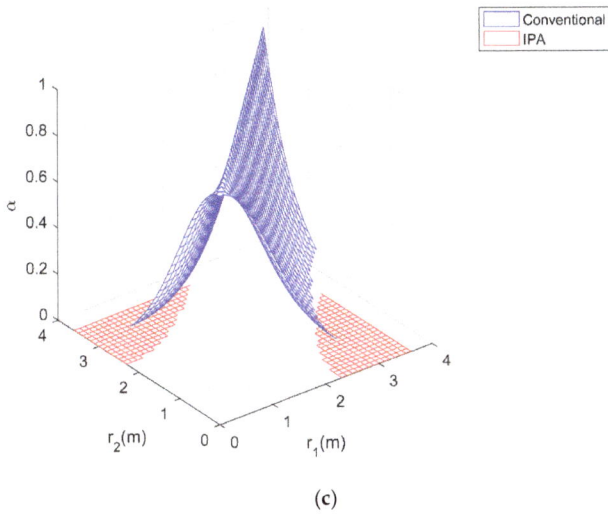

(c)

Figure 5. Simulation results for Scenario 1 based on the proposed adaptive power allocation scheme, gain ratio power allocation (GRPA) scheme, and IPA scheme, respectively: (**a**) total achievable data rate; (**b**) user fairness; (**c**) optimal power allocation factor.

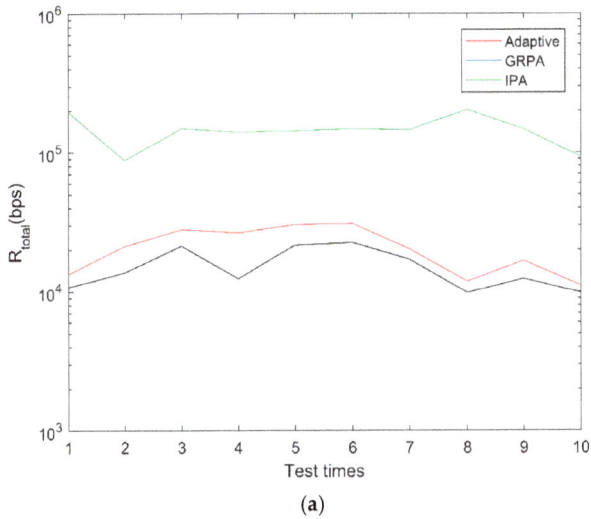

(a)

Figure 6. *Cont.*

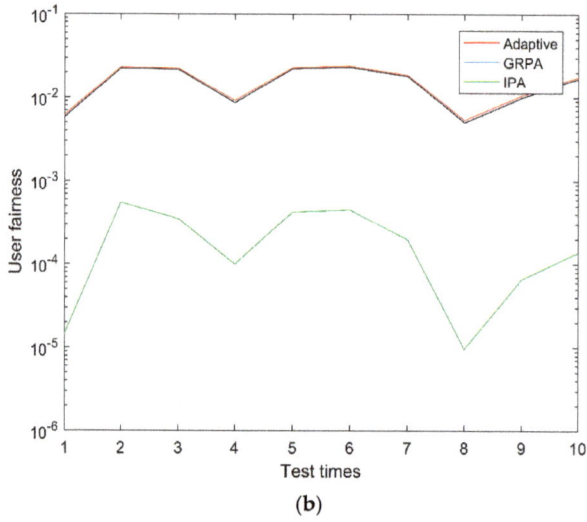

(b)

Figure 6. Simulation results for Scenario 2 based on the proposed adaptive power allocation scheme, GRPA scheme, and IPA scheme, respectively, wherein the randomly generated combinations of $r_1, r_2, \cdots, r_5$ for the 10 tests are (3.5309, 2.8755, 2.2167, 0.5458, 0.3096), (3.4397, 2.4176, 2.3547, 1.6103, 1.4316), (3.0480, 2.3202, 1.7318, 1.3096, 0.8124), (3.5805, 2.1633, 2.1595, 1.9316, 0.9357), (3.0654, 2.2295, 1.4686, 1.4371, 0.8664), (3.0212, 2.2223, 1.5741, 1.1445, 0.8263), (3.1828, 2.5616, 2.4763, 2.1451, 0.8921), (3.5729, 2.9899, 2.1466, 0.7575, 0.1092), (3.4661, 2.6299, 2.5681, 1.3844, 0.8624), and (3.5628, 3.0949, 2.8034, 2.5635, 1.4087), respectively: (**a**) total achievable data rate; (**b**) user fairness.

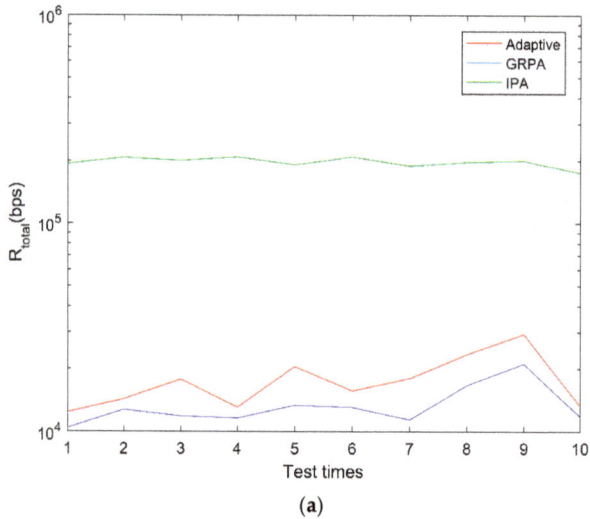

(a)

Figure 7. *Cont.*

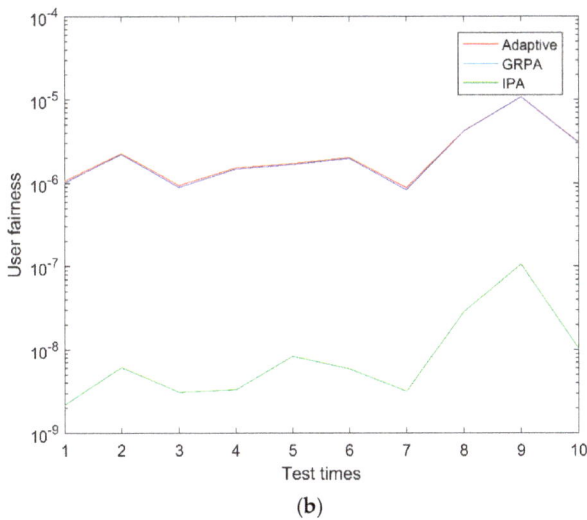

(b)

Figure 7. Simulation results for Scenario 3 based on the proposed adaptive power allocation scheme, GRPA scheme, and IPA scheme, respectively, wherein the randomly generated combinations of $r_1, r_2, \cdots, r_{10}$ for the 10 tests are (3.5635, 2.9895, 2.6482, 2.2378, 2.1181, 1.2470, 1.0532, 0.9620, 0.9131, 0.3908), (3.3684, 2.9048, 2.6485, 2.0980, 1.7719, 1.6640, 0.9519, 0.8605, 0.3101, 0.1060), (3.5600, 2.5837, 2.4566, 2.1188, 1.6979, 1.6005, 1.0897, 0.9554, 0.6994, 0.2629), (3.4459, 2.9862, 2.5604, 2.5197, 1.1499, 1.0042, 0.3522, 0.1674, 0.1249, 0.1154), (3.4804, 2.4649, 2.3756, 2.1224, 1.8073, 1.2344, 1.0010, 0.8116, 0.5897, 0.4315), (3.3871, 2.8256, 2.0628, 1.9249, 1.7022, 1.2225, 0.4711, 0.2751, 0.1956, 0.0432), (3.6230, 2.5424, 2.5316, 2.4171, 2.4164, 1.9551, 1.0932, 0.6460, 0.6205, 0.4642), (3.2883, 2.3745, 2.3405, 2.1248, 1.4024, 1.0412, 0.6931, 0.4675, 0.4042, 0.3604), (3.0934, 2.2558, 1.8612, 1.4571, 1.2727, 0.8702, 0.8700, 0.6669, 0.4472, 0.2755), and (3.4347, 2.9254, 2.8023, 2.7768, 2.0362, 1.7961, 1.4943, 1.2546, 1.0903, 0.6240), respectively: (**a**) total achievable data rate; (**b**) user fairness.

For Scenario 1, with regard to the proposed adaptive power allocation scheme, we can conclude from Figure 5a,b that when the distance between the two users is large enough, namely, when one is near the center of the optical attocell and the other one is near the edge, the total achievable data rate approaches a maximum of about 2.5×10^5 bps and the user fairness approaches a minimum of about zero. This is because the IPA scheme, which corresponds to the red region illustrated in Figure 5c, is adopted in this case. Note that this case is less common than the one in which the conventional power allocation scheme is adopted and that the value of the total achievable data rate and user fairness varies smoothly between 0.3×10^5 bps and 1.2×10^5 bps, and between 0.9 and 1, respectively. In addition, when comparing the proposed scheme with the GRPA scheme, we find that the total achievable data rate increases greatly at a small cost of user fairness; when comparing the proposed scheme with the IPA scheme, we find that the user fairness is improved greatly at a small cost of total achievable data rate. Specifically, when the proposed scheme, GRPA scheme, and IPA scheme are adopted, the mean values of the total achievable data rate are 9.7748×10^4 bps, 5.8142×10^4 bps, and 1.1455×10^5 bps, respectively, and the mean values of user fairness are 0.6913, 0.9989, and 0.2382, respectively. Apparently, the total achievable data rate gain of the proposed scheme reaches 68.2% compared to that of the GRPA scheme, while user fairness is reduced by only 30.8%; the user fairness gain of the proposed scheme reaches 190.2% compared with that of the IPA scheme, while the total achievable data rate is reduced by only 14.7%. Hence, we can conclude that the proposed adaptive scheme facilitates a better balance between total achievable data rate and user fairness. Moreover, from Figure 5c, we find that no matter which power allocation scheme is adopted, the optimal power allocation factor follows these two rules: first, the optimal power allocation factor increases with decreasing distance between the two users; and second, when the distance between the two users remains unchanged, the optimal power allocation factor increases with an increase in the mean r_1 and r_2.

For Scenario 2, by considering Figure 6, we can easily obtain the mean values of the total achievable data rate and user fairness based on the three schemes. Specifically, compared with the GRPA scheme, the proposed scheme increases the total achievable data rate by about 38.51% at no cost of user fairness; compared with the IPA scheme, the proposed scheme increases the user fairness by a factor of 66.6750 and reduces the total achievable data rate by only 0.8556. Similarly, for Scenario 3, we can conclude from Figure 7 that the proposed scheme increases the total achievable data rate by about 32.54% at no cost of user fairness compared with the GRPA scheme and improves the user fairness by a factor of 156.6884 with a 90.97% loss of the total achievable data rate compared with the IPA scheme. When considering Figures 5–7 comprehensively, we find that the performance gain of the proposed scheme is relatively considerable, even with the increase in the number of users. The feasibility and superiority of our proposed adaptive scheme is verified accordingly.

It should be emphasized that our proposed adaptive power allocation scheme is not only suitable for the situation in which the IPA scheme is involved. If we aim to perform NOMA without sacrificing user fairness, that is to say, if only the conventional power allocation scheme is considered, our proposed adaptive scheme can also play a significant role in exploring the optimal power allocation factor based on the real time location of users and the MADM algorithm.

6. Modeling of Optimal Power Allocation Factor for Mobile NOMA-VLC

In order to further study the change rule of the optimal power allocation factor for mobile NOMA-VLC systems, we consider a mobile scenario, as shown in Figure 8, and establish the fitting model of the optimal power allocation factor in that case, in which two users walk along Trajectory 1 and Trajectory 2, respectively, with the same velocity (see Figure 8 for blue and red lines). Other parameters are the same as in Table 1.

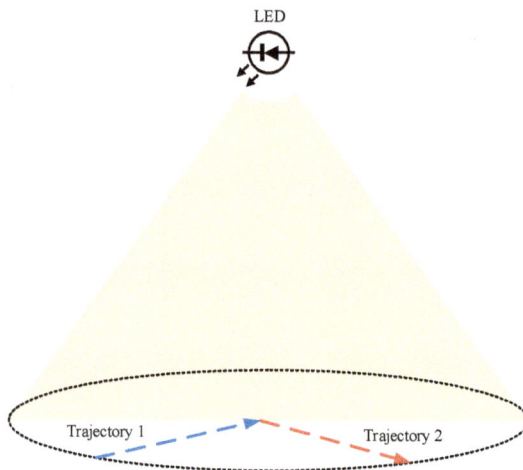

Figure 8. Movement trajectories of User 1 and User 2.

Based on our proposed adaptive power allocation scheme, we first obtain the optimal power allocation factor along with the movement distance of the two users, in which the simulation step is still set to 0.1 m. Then, we apply curve fitting techniques to establish the fitting model, in which the "leave-one-out cross-validation (LOOCV)" method [23] is adopted to avoid over-fitting. Specifically, we take four functions into account for curve fitting, i.e., exponential, Fourier, Gaussian, and sinusoidal.

In addition, in order to guarantee the conciseness and effectiveness of the fitting model, we limit the number of its terms to three or less, thus yielding 12 tests. Moreover, a nonlinear least squares (NLS) method is adopted due to its intrinsic capability to fit a large range of functions and to produce good estimates of the unknown parameters from small data sets. Based on a trust-region algorithm, this method tries to refine the parameters by iterative optimization, and we set the maximum number of iterations to 400. In the process, we used the root mean square error (RMSE) and R-square to assess the fitting accuracy; a better fitting model possesses a smaller value of RMSE and a value of R-square closer to 1. The best curve fitting is illustrated in Figure 9, and the corresponding RMSE value and R-square value are 0.01627 and 0.9953, respectively.

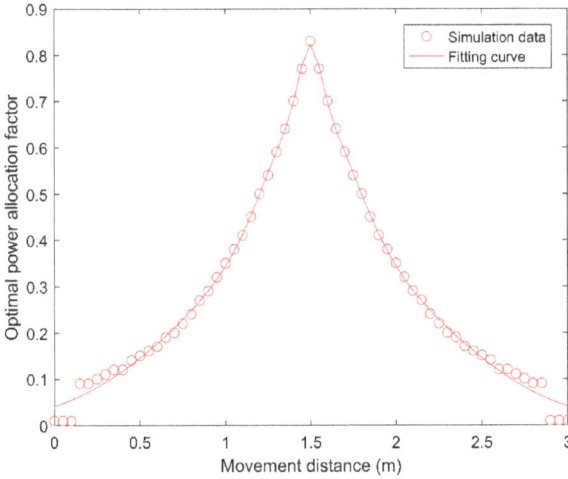

Figure 9. Optimal power allocation factor versus movement distance.

Moreover, according to the curve fitting, the fitting model can be expressed as

$$\alpha_{\text{optimal}} = \sum_{j=1}^{3} l_j \exp(-((\zeta - m_j)/n_j)2), \tag{26}$$

where α_{optimal} denotes the optimal power allocation factor; ζ denotes the movement distance of users; and the related coefficients l_j, m_j, and n_j are presented in Table 2.

Table 2. Coefficients used in Equation (26).

Coefficient	Value	Coefficient	Value	Coefficient	Value
l_1	0.1299	m_1	1.5	n_1	0.08495
l_2	0.2906	m_2	1.5	n_2	0.3375
l_3	0.4018	m_3	1.5	n_3	0.99

Next, we extend the model of the optimal power allocation factor to adapt to more realistic scenarios in which the transmitting power P_{elec} of the LED can be tunable. First, through simulations, we observe that under different values of P_{elec}, the variation of α_{optimal} along with movement distance may always be approximated as the sum of Gaussian functions as before, but with some shifting along the y-axis. Hence, we assume the extended model as follows:

$$\alpha_{\text{optimal}} = f(P_{\text{elec}}) + \sum_{j=1}^{3} l_j \exp(-((\zeta - m_j)/n_j)2). \tag{27}$$

In order to determine which functional form $f(P_{\text{elec}})$ obeys and the corresponding coefficients, we took the polynomial, exponential, and Gaussian function forms into account and created a function nlinfit() in MATLAB R2016a to perform nonlinear regression based on a data set. The resulting best-fitting model with an RMSE of 0.029 was

$$\alpha_{\text{optimal}} = t_1 + t_2 P_{\text{elec}} + t_3 P_{\text{elec}}^2 + t_4 P_{\text{elec}}^3 + \sum_{j=1}^{3} l_j \exp(-((\zeta - m_j)/n_j)2), \tag{28}$$

where the related coefficients $t_1, t_2, t_3, t_4, l_j, m_j$, and n_j are presented in Table 3.

Table 3. Coefficients in Equation (28).

Coefficient	Value
t_1	-0.0223
t_2	7.7957
t_3	-2.3387×10^3
t_4	1.8817×10^5
l_1	0.1249
l_2	0.2581
l_3	0.3609
m_1	1.5
m_2	1.5
m_3	1.5
n_1	0.0999
n_2	0.4032
n_3	1.1839

In order to verify the model's effectiveness under different P_{elec} values, we conducted a simulation, with the result shown in Figure 10.

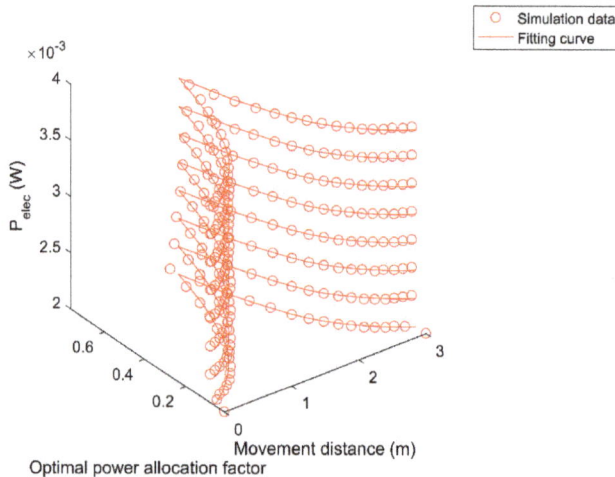

Figure 10. Optimal power allocation factor versus movement distance under different P_{elec} values.

7. Conclusions

In this paper, we put forward a novel concept named the "inverse power allocation (IPA)" which can lead to a higher total achievable data rate compared to conventional power allocation systems, at a cost of user fairness. Then, we proposed an adaptive power allocation scheme based on a MADM

Electronics **2019**, *8*, 381

algorithm, in which the total achievable data rate and user fairness were considered comprehensively via mathematical assessment; through simulation, the conditions under which the IPA scheme or the conventional scheme will be adopted were observed and the optimal power allocation factor was obtained according to users' locations. Finally, after assuming users walk along certain trajectories, we studied a variation model of the optimal power allocation factor along with users' movement distances utilizing the curve fitting technique and derived a fitting model.

Author Contributions: Conceptualization, Z.D. and T.S.; formal analysis, Z.D. and T.S.; funding acquisition, T.S.; investigation, Q.L. and T.T.; methodology, Z.D. and T.T.; software, Z.D. and T.T.; validation, T.S. and Q.L.; writing—original draft, Z.D. and Q.L.; writing—review and editing, Z.D. and T.S.

Funding: This research was funded by the National Natural Science Foundation of China, grant numbers 61771357 and 61172080.

Conflicts of Interest: The authors declare no conflict of interest. The funders had no role in the design of the study; in the collection, analyses, or interpretation of data; in the writing of the manuscript, or in the decision to publish the results.

References

1. Andrews, J.G.; Buzzi, S.; Wan, C.; Hanly, S.V.; Lozano, A.; Soong, A.C.K. What will 5g be? *IEEE J. Sel. Areas Commun.* **2014**, *32*, 1065–1082. [CrossRef]
2. Jovicic, A.; Li, J.; Richardson, T. Visible light communication: Opportunities, challenges and the path to market. *IEEE Commun. Mag.* **2013**, *51*, 26–32. [CrossRef]
3. Burchardt, H.; Serafimovski, N.; Tsonev, D.; Videv, S.; Haas, H. VLC: Beyond point-to-point communication. *IEEE Commun. Mag.* **2014**, *52*, 98–105. [CrossRef]
4. Haas, H.; Yin, L.; Wang, Y.; Chen, C. What is LiFi? *IEEE J. Lightw. Technol.* **2016**, *34*, 1533–1544. [CrossRef]
5. Pathak, P.H.; Feng, X.; Hu, P.; Mohapatra, P. Visible light communication, networking, and sensing: A survey, potential and challenges. *IEEE Commun. Surv. Tutur.* **2015**, *17*, 2047–2077. [CrossRef]
6. Saito, Y.; Kishiyama, Y.; Benjebbour, A.; Nakamura, T.; Li, A.; Higuchi, K. Non-orthogonal multiple access (NOMA) for cellular future radio access. In Proceedings of the 2013 IEEE 77th Vehicular Technology Conference (VTC Spring), Dresden, Germany, 2–5 June 2013; pp. 1–5. [CrossRef]
7. Saito, Y.; Benjebbour, A.; Kishiyama, Y.; Nakamura, T. System level performance evaluation of downlink non-orthogonal multiple access (NOMA). In Proceedings of the 2013 IEEE 24th Annual International Symposium on Personal, Indoor, and Mobile Radio Communications (PIMRC), London, UK, 8–9 September 2013; pp. 611–615. [CrossRef]
8. Ding, Z.; Yang, Z.; Fan, P.; Poor, H.V. On the performance of non-orthogonal multiple access in 5G systems with randomly deployed users. *IEEE Signal Process. Lett.* **2014**, *21*, 1501–1505. [CrossRef]
9. Marshoud, H.; Kapinas, V.M.; Karagiannidis, G.K.; Muhaidat, S. Non-orthogonal multiple access for visible light communications. *IEEE Photonics Technol. Lett.* **2016**, *28*, 51–54. [CrossRef]
10. Yin, L.; Popoola, W.O.; Wu, X.; Haas, H. Performance evaluation of non-orthogonal multiple access in visible light communication. *IEEE Trans. Commun.* **2016**, *64*, 5162–5175. [CrossRef]
11. Guan, X.; Yang, Q.; Hong, Y.; Chan, C.C.K. Non-orthogonal multiple access with phase pre-distortion in visible light communication. *Opt. Express* **2016**, *24*, 25816–25823. [CrossRef] [PubMed]
12. Fu, Y.; Hong, Y.; Chen, L.K.; Sung, C.W. Enhanced power allocation for sum rate maximization in OFDM-NOMA VLC systems. *IEEE Photonics Technol. Lett.* **2018**, *30*, 1218–1221. [CrossRef]
13. Chen, C.; Zhong, W.D.; Yang, H.; Du, P. On the performance of MIMO-NOMA based visible light communication systems. *IEEE Photonics Technol. Lett.* **2018**, *30*, 307–310. [CrossRef]
14. Yin, L.; Wu, X.; Haas, H. On the performance of non-orthogonal multiple access in visible light communication. In Proceedings of the 2015 26th Annual Symposium on Personal, Indoor, and Mobile Radio Communications (PIMRC), Hong Kong, China, 30 August–2 September 2015; pp. 1354–1359. [CrossRef]
15. Zhang, X.; Gao, Q.; Gong, C.; Xu, Z. User grouping and power allocation for NOMA visible light communication multi-cell networks. *IEEE Commun. Lett.* **2017**, *21*, 777–780. [CrossRef]
16. Zeng, L.; O'Brien, D.C.; Minh, H.L.; Faulkner, G.E.; Lee, K.; Jung, D. High data rate multiple input multiple output (MIMO) optical wireless communications using white led lighting. *IEEE J. Sel. Areas Commun.* **2009**, *27*, 1654–1662. [CrossRef]

17. Kahn, J.M.; Barry, J.R. Wireless infrared communications. *Proc. IEEE* **1997**, *85*, 265–298. [CrossRef]
18. Andrews, J.G.; Meng, T.H. Optimum power control for successive interference cancellation with imperfect channel estimation. *IEEE Trans. Wirel. Commun.* **2003**, *2*, 375–383. [CrossRef]
19. Datasheet of BPW21R. Available online: https://pdf1.alldatasheetcn.com/datasheet-pdf/view/26249/VISHAY/BPW21R.html (accessed on 20 March 2019).
20. Wang, Y.M. A method based on standard and mean deviations for determining the weight coefficients of multiple attributes and its applications. *Appl. Stat. Manag.* **2003**, *22*, 22–26. [CrossRef]
21. Sheng-Mei, L.; Su, P.; Ming-Hai, X. An improved TOPSIS vertical handoff algorithm for heterogeneous wireless networks. In Proceedings of the 2010 IEEE 12th International Conference on Communication Technology (ICCT 2010), Nanjing, China, 11–14 November 2010; pp. 750–754. [CrossRef]
22. Lahby, M.; Cherkaoui, L.; Adib, A. Performance analysis of normalization techniques for network selection access in heterogeneous wireless networks. In Proceedings of the 2014 IEEE 9th International Conference on Intelligent Systems: Theories and Applications (SITA-14), Rabat, Morocco, 7–8 May 2014; pp. 1–5. [CrossRef]
23. James, G.; Witten, D.; Hastie, T.; Tibshirani, R. *An Introduction to Statistical Learning;* Springer: New York, NY, USA, 2013.

electronics

MDPI

Article

Color-Independent Visible Light Communications Based on Color Space: State of the Art and Potentials

Tae-Ho Kwon [1], Jai-Eun Kim [1], Youn-Hee Kim [2] and Ki-Doo Kim [1,*]

[1] School of Electronics Engineering, Kookmin University, Seoul 02707, Korea;
    kmjkth@kookmin.ac.kr (T.-H.K.); eun9477@kookmin.ac.kr (J.-E.K.)
[2] Department of Fashion Design, Kookmin University, Seoul 02707, Korea; shell62@kookmin.ac.kr
*   Correspondence: kdk@kookmin.ac.kr; Tel.: +82-2-910-4707

Received: 23 July 2018; Accepted: 7 September 2018; Published: 10 September 2018

Abstract: Color independency is an important factor in visible light communication (VLC) systems. This paper aims to review and summarize recent achievements in color-independent visible light communication based on color space, with the main focus being on color-space-based modulation (CSBM), termed as generalized color modulation (GCM), which allows VLC to adapt to any target color. The main advantages of GCM are its color independency, reasonable bit error rate (BER) performance during color variation, and dimming control. We also address our past research works that aimed to achieve a color-independent visual MIMO system by incorporating the advantages of GCM, which can lead to higher data rates over longer distances and improved performance, using image processing in addition to color independency. Finally, two case studies are introduced to demonstrate the potential applicability of a color-independent visual-MIMO system using color-space-based modulation techniques.

Keywords: visible light communication; generalized color modulation; color-space-based modulation; color independence; visual MIMO; wearable device; V2X

1. Introduction

Visible light communication (VLC) has recently become popular as an alternative to radio and infrared communication systems [1–3]. Visible light has several features that distinguish it from radio frequencies, such as the availability of unique bandwidths that are not subject to interference from electromagnetic waves. At the same time, light-emitting diodes (LEDs) offer benefits over other light sources, such as high speed, small size, long lifetime, and rapid switching. By combining these two factors, it is possible to provide high-speed communication using VLC while maintaining the basic functionality (e.g., illumination, displays, etc.) of visible light.

Compared to radio frequency (RF) communication systems, VLC systems offer distinct channel characteristics and signal sources. Variable on-off keying (VOOK), variable pulse position modulation (VPPM), multiple PPM (MPPM), pulse dual slope modulation (PDSM), orthogonal frequency division multiplexing (OFDM), and subcarrier modulations are well-known modulation schemes of VLC systems [4–8]. Note that intensity variation of the optical signal is the main key point of these modulation schemes. In wavelength division multiplexing (WDM), different wavelengths are used to multiplex various optical signals into a single optical fiber [9]. WDM can be used in VLC systems by implementing it with LEDs of different wavelengths by varying the intensity of the signals to transmit information data. To receive the transmitted data, a photodiode (PD) along with different band-pass filters are used. Finally, the transmitted data is recovered by demodulating the received signal on PD.

Color-shift keying (CSK), the first color space-based modulation scheme, was proposed by the IEEE 802.15.7 Task Group [10,11]. However, CSK is not suitable for communication under target color variation and, in which case, CSK may exhibit lower performance than traditional intensity-based

WDM [10]. The target color indicates the desired color for LED lighting. In a past paper [12], constellation designs for CSK were examined using billiard algorithms, however, any analytical verification could not be shown for the solution of constellation design for balancing color. In numerous instances of the literature [13–16], to overcome this limitation, generalized color modulation (GCM), one of the color-space-based modulation (CSBM) methods, was introduced and resolved for the color-independent VLC systems. The term "color-independent" implies the independence of light color and intensity variations. In this way, a VLC system can perform seamless communication that can preserve the originality and brightness of lighting color. In other words, GCM allows communication without compromising the original lighting color. As a result, any data stream can be delivered using GCM regardless of the target color. GCM also has a number of other advantageous features, such as dimming control, color independence, independence from the number of LEDs and PDs, and acceptable BER performance in the presence of color variation. Figure 1 presents a simple conceptual block diagram of a color-space-based VLC system in comparison to a carrier-signal-based RF system.

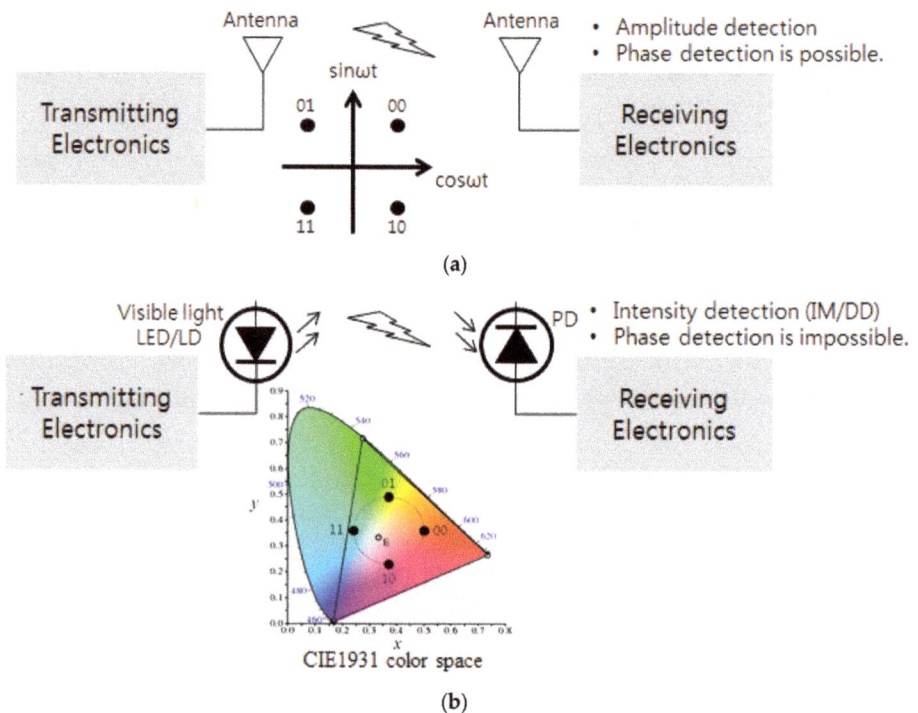

Figure 1. Color-space-based visible light communication (VLC) system vs. carrier signal-based RF system. (**a**) RF communication with carrier-based constellation; (**b**) VLC with color-space-based constellation.

However, previous studies in wireless optical communication using visible light with photodiode receivers have either been limited to short distances or have required complicated processing at the receiver. Photodiodes can convert pulses at a fairly high rate but suffer from significant interference and background light noise. As a result, the signal-to-noise ratio (SNR) is very low and the communication range shortened. It is possible to overcome this data transmission rate limitation and thus realize a larger transmission range by employing a camera as a receiver and a light emitting array as a transmitter, based on visual multiple input multiple output (MIMO) [17,18].

2. Light Color Spaces

Colors can be represented by various forms of color space, such as CIE1931 (CIEXYZ), CIE1976 (CIELUV), RGB, HSV, HSL, CMY, YUV (PAL), and YIQ (NTSC) [19]. In the color space, sources (e.g., LEDs) and receiver (e.g., PDs) can be denoted as points. The minimum geometric area of a color space that contains all LED points is defined as the gamut area. Inside this area, any color can be created by combining the colors from the LEDs. The following two types of color space are particularly useful for the system proposed in this paper.

2.1. Color Space CIE1931

We consider the CIE1931 color space because of its simplicity. The tristimulus values (X, Y, and Z) represent the three primary color quantities of the three-component additive color model needed to produce a target color [19].

$$X = \int S(\lambda)\,\bar{x}(\lambda)d\lambda, \ Y = \int S(\lambda)\,\bar{y}(\lambda)d\lambda, \ Z = \int S(\lambda)\,\bar{z}(\lambda)d\lambda \tag{1}$$

In (1), λ indicates the wavelength (nanometers) of the monochromatic light for each corresponding primary color. $\bar{x}(\lambda)$, $\bar{y}(\lambda)$, and $\bar{z}(\lambda)$ are the three color matching functions for the CIE1931 color space, and $S(\lambda)$ is the spectral power distribution of a light source. The three normalized tristimulus values can be determined by

$$x = \frac{X}{X+Y+Z}, \ y = \frac{Y}{X+Y+Z}, \ z = \frac{Z}{X+Y+Z} \tag{2}$$

$$x + y + z = 1 \tag{3}$$

Using these normalized values, a CIE1931 color space chromaticity diagram can be produced [19]. The xy-plane of a chromaticity diagram is shown in Figure 2. The CIE1931 chromaticity diagram is not perceptually uniform since it is derived from the human-eye response function.

To determine the color triple (X, Y, Z) from the (x,y) coordinates of the CIE1931 color space, information for Y (luminance) must be known. We can calculate the tristimulus values using (2) and (3) [19].

$$z = 1 - x - y, \ X = \frac{x}{y}Y, \ Z = \frac{z}{y}Y \tag{4}$$

The (x, y) coordinates represent the colors in the CIE1931 color space. The colors can also be represented by the proportion of R, G, and B in the RGB color space; the wavelength of a monochromatic light source in these three channels is 700 nm, 546 nm, and 436 nm, respectively. The XYZ and RGB coordinate systems have a linear relationship with each other, thus one can be transformed into the other [19].

$$\begin{bmatrix} X \\ Y \\ Z \end{bmatrix} = \frac{1}{0.17697} \begin{bmatrix} 0.4900 & 0.3100 & 0.2000 \\ 0.17697 & 0.81240 & 0.01063 \\ 0.0000 & 0.0100 & 0.9900 \end{bmatrix} \begin{bmatrix} R \\ G \\ B \end{bmatrix} \tag{5}$$

$$\begin{bmatrix} R \\ G \\ B \end{bmatrix} = \begin{bmatrix} 0.4185 & -0.1587 & -0.0828 \\ -0.0812 & 0.2524 & 0.0157 \\ 0.0009 & -0.0025 & 0.1786 \end{bmatrix} \begin{bmatrix} X \\ Y \\ Z \end{bmatrix} \tag{6}$$

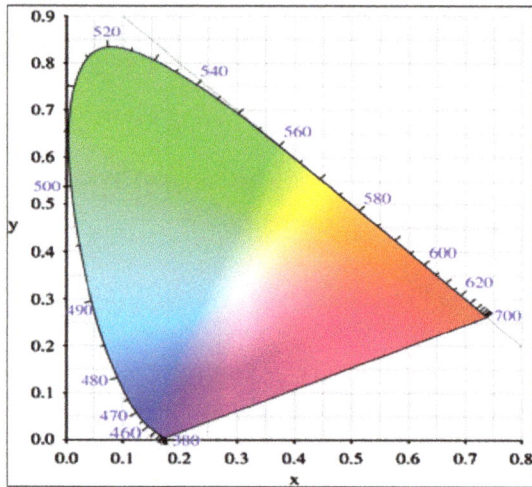

Figure 2. CIE1931 color space chromaticity diagram [15].

2.2. Color Space CIE L*u*v* (CIELUV)

A detailed description of the CIELUV color space is included in a past paper [19]. The key advantage of this color space is that the distance between two points is approximately proportional to the perceived color difference. The color resulting from the addition of two different colors will fall on a connecting line. CIELUV has better uniformity for perceived colors than does CIE1931. Therefore, our proposed modulation schemes [13,14] are based on the CIELUV color space in order to have perceptual uniformity that is close to that of the human eye in the transmitter (LEDs) and receiver (PDs).

3. Generalized Color Modulation

GCM was introduced to make the VLC system color-independent [13,14]. Using this scheme, it is possible to send any data stream with multiple LEDs that are independent of the target color of the VLC system. Color independency is one of the most important advantages of GCM over other VLC modulation schemes. Because GCM can produce any target color within a gamut area by combining other colors, it is possible to establish accurate communication using a VLC system while maintaining the original color and brightness. The target color can also be demapped on the receiver side from the received color symbols without having information about the target color. Generally, in WDM, transmitter and receiver use a definite number of wavelengths. In contrast, in GCM, the number of wavelengths can be chosen within the visible band depending on the application. In other words, GCM is independent of the number of LEDs and PDs. In addition, a fixed total light intensity can be maintained over time, which allows dimming control and flicker-free operation, whereas other intensity-based modulations (e.g., OOK and PPM) alter the light intensity to send data.

To represent the target color, GCM uses a constellation diagram with 2^m points (m-bit data) in a color space, with each point in this diagram representing a transmitted data symbol. They are mapped onto RGB LEDs using (4) and (6). On the receiver side, PDs detect the RGB intensities, from which the (x, y) coordinates are determined using (5) and (2). Finally, the error boundary decision (EBD) and the Euclidean distance are used to demap the received points into the original data symbols [15].

Because OFDM is known to be a useful modulation method for VLC systems for several reasons [7,20,21], an OFDM-based VLC system that can be color independent using CSBM was presented previously [22]. With all the promising advantages of OFDM, the proposed system can be applied to all colors in the visible band. In a past paper [22], it demonstrated robustness of the proposed OFDM-VLC system to inter-symbol-interference (ISI) and a large peak-to-average power ratio (PAPR) while maintaining color independency.

In GCM, we define the data symbols using a constellation diagram in the color space. We generate the constellation based on two assumptions:

- The transmitted data symbols are selected in such a way that they are random and equidistant from each other on the circumference of the constellation diagram.
- The changing rates of the target color are lower than the data rates.

To generate the constellation diagram, we consider two aspects when determining the points in the light color space: (i) colors as symbols are used to determine a target color perceivable to the human eye and (ii) the maximum distance between two adjacent constellation points are preferred in the constellation diagram. In order to reduce the symbol error rate (SER), the first assumption is the best choice because it minimizes the effect of interference. Additionally, the area of the constellation diagram has to be maximized when the second aspect is considered.

The area of constellation diagram is maximized in two ways: (i) the coordinate of the target color becomes the center of constellation diagram and (ii) the coordinates of the LEDs are used to form the gamut area. Based on this, the maximum area of a constellation diagram is constructed by drawing the largest circle inside the gamut area in a manner so that the center of the largest circle represents the target color point. Figure 3 represents the generation of a constellation diagram with a target color [14]. In this case, the constellation points are arranged using a similar arrangement to that of RF circular quadrature amplitude modulation (QAM). More examples of circle-type and line-type constellation diagrams can be found in a past paper [16]. We suppose that equiprobable symbol transmission is valid due to the compensation and interleaving algorithm. The target color can then be obtained from the averaged RGB value with a number of symbols as described in (7),

$$(x_t, y_t) = \left(\frac{\sum_{i=1}^{N} x_i}{N}, \frac{\sum_{i=1}^{N} y_i}{N} \right), \tag{7}$$

where (x_t, y_t) denotes the position of the target color, (x_i, y_i) denotes the position of the i_{th} symbol, and N is the number of symbols used to calculate the moving average. From the sense of probability, as N increases, (x_t, y_t) moves closer to the actual target color.

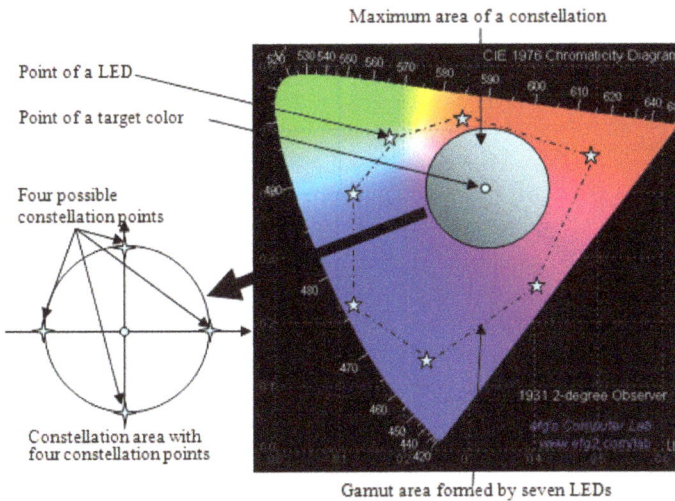

Figure 3. Generation of a constellation diagram in Color Space CIE L*u*v* (CIELUV) color space [14].

4. CSBM-Based VLC System Description

Figure 4 presents the total representation of a CSBM-based VLC system [13,16]. After converting the input from serial to parallel, we define the data symbols from the light color space using constellation points. A point can be mapped to the intensities of n LEDs at the transmitter. The average of every constellation point becomes the target color. The target color may be selected from the gamut area, and the information data can be transmitted using constellation points corresponding to the target color. Thus, the proposed system enables color-independent visible light communication. On the receiving side, the target color information is optional because it can be generated from the average of received data symbols [14]. After that, intensities at the PDs are amplified. Finally, after the demapping operation, m-bit output is converted into a serial data stream.

A detailed explanation of mapping and demapping can be found in the literature [13,16].

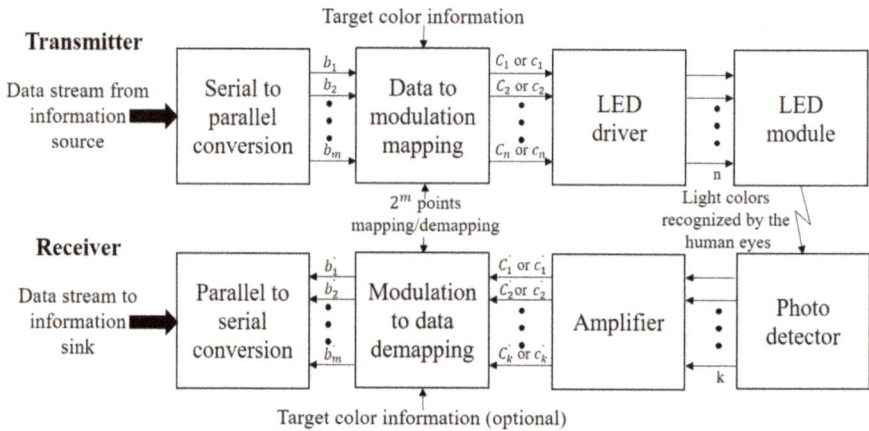

Figure 4. System representation of a color-space-based modulation (CSBM)-based VLC [13].

Additive white Gaussian noise (AWGN) is assumed as channel noise. Typically, it is caused by background lighting that can interfere with the VLC signal. Optical filters can be useful for minimizing the noise, but even if the receiver is well designed, there may still be shot noise. On the receiver side, three PDs with three RGB filters were used to detect the light. To obtain the target color on the receiver side, the following cases are considered [15].

- Case-I (sending the target color): We add the target color information with the transmitted data symbols using the LED intensities as header information, thus differentiating it from the transmitted data symbols.
- Case II (without sending the target color): By taking a moving average of the received symbols, we can determine the target color [14].

If the transmitter does not send the target color, many errors can occur while switching from one target color to another. This looks similar to burst errors of RF communication because of the appearance of a sequence of successive errors in the target color.

Figure 5 presents the BER performance of CSBM for both Case-I and Case-II, with and without channel coding, respectively [15]. We can see that, unlike in Case I, the BER performance of Case II deteriorates without channel coding because of the burst errors occurring during the transition period when the transmitter does not send the target color. However, with channel coding, BER performance is improved by minimizing the random and burst errors, and the improved result is almost similar to that of Case-I.

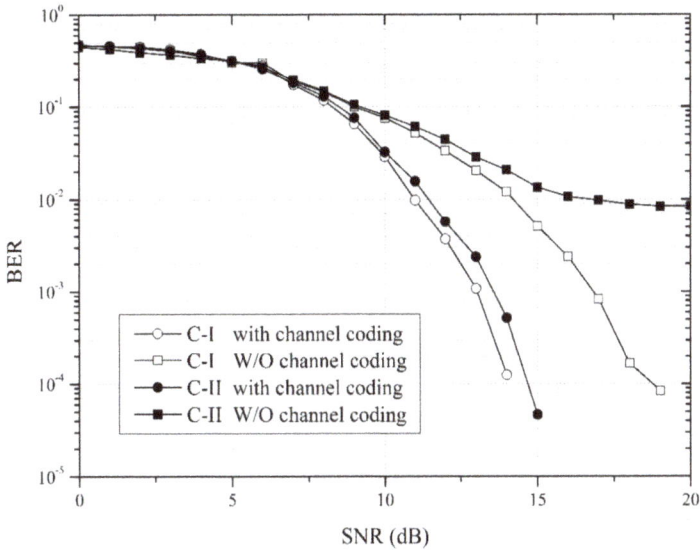

Figure 5. Bit error rate (BER) versus SNR for Case-I and Case-II (with and without channel coding) [15].

We used MATLAB to simulate our proposed system. The simulation model is represented in Figure 6 [14]. First, a built-in MATLAB function was used to generate random data. FEC coding (1/2 convolutional coding) and interleaving were then adopted for the input data bits. Following GCM mapping, the data bits were mapped onto LED modules. Because AWGN is used as noise in our model, it was added to the transmitted data. We model the AWGN depending on channel SNR using MATLAB. On the receiver side, deinterleaving and Viterbi decoding were used after GCM demapping to retrieve the data. The Euclidean distance was used for symbol decision during the GCM demapping process. Finally, the received data is compared with the transmitted data to calculate the BER.

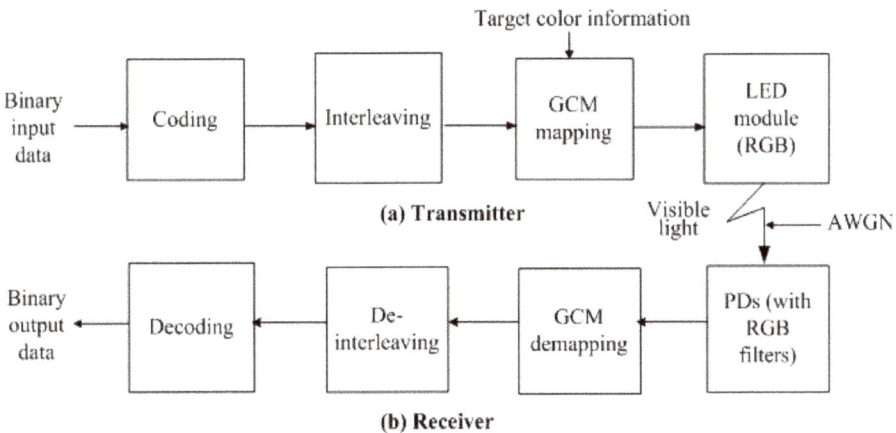

(a) Transmitter

(b) Receiver

Figure 6. Simple simulation model used to demonstrate the proposed generalized color modulation (GCM)-based VLC system [14].

Figure 7 presents the generation of three target colors for the transmitter and their corresponding symbols in the CIELUV color space [13,14]. Figure 8 displays the BER performance for the three single target colors [13,14]. BER performance differs noticeably because of the different sizes of the constellation diagrams for each target color.

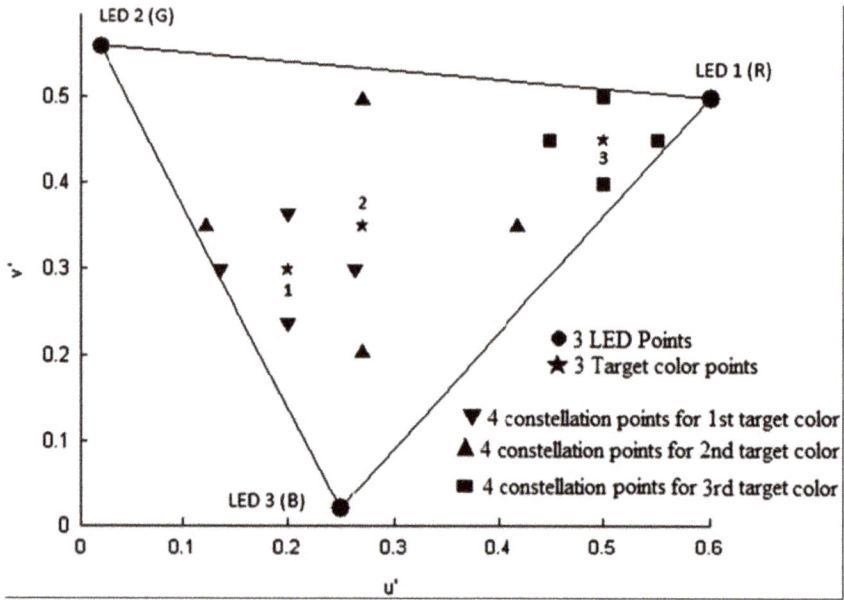

Figure 7. Generation of constellations points corresponding to three target colors in the CIELUV color space [13].

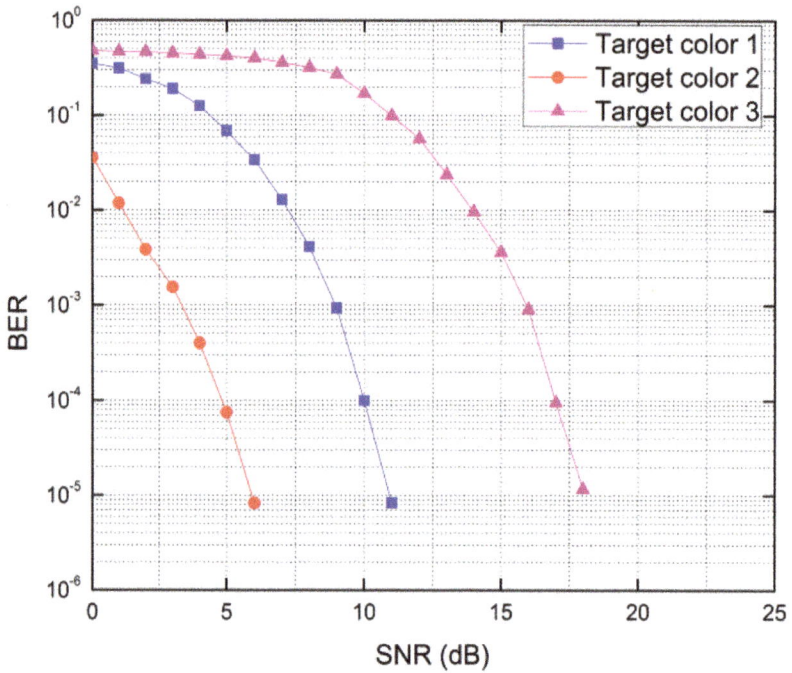

Figure 8. Comparison of BER vs. SNR for three target colors [13].

5. Visual MIMO with Color-Space-Based Modulation

In a past paper [17], it was shown that higher data rates are possible in long-range transmissions for mobile optical communications by a camera using the concept "visual-MIMO". According to the concept, the optical transmission of a light-emitting array is received by the photodetector elements (i.e., pixels) of a camera. The image sensor of a camera consists of these pixels, defined as an array of receiver which is inherently highly directional. This system provides a degree of freedom in selecting and combining a subset of receiver elements that receive a strong signal from the transmitter and have a large SNR. Conceptually, this visual MIMO system may be quite similar to RF-MIMO antenna selection, but visual-MIMO has less overhead and less complexity to process at the camera receiver due to the use of image processing and computer vision algorithms [17].

Figure 9 presents the color-independent visual-MIMO communication system [23,24]. GCM-based visual-MIMO helps to achieve better performance in terms of SER than conventional light-emitting diode (LED) communication. Using this scheme, it is possible to transmit different colors (symbols) at the same time through LED array. Most importantly, the proposed visual -MIMO system can easily adapt to changes in the target color via image processing [23,24].

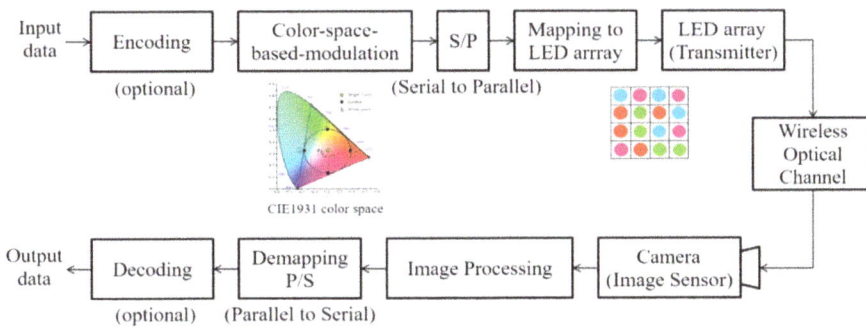

Figure 9. Color-independent visual MIMO transceiving process [23].

In general, the capacity of a color-independent visual MIMO system depends on the number of bits per symbol, the array size (M × M), and the frame rate of the camera. The capacity of the proposed system is defined as,

$$\text{Capacity [bps]} = n \text{ [bps/symbol]} \times N \text{ [symbols/frame]} \times F \text{ [frames/ sec]}, \tag{8}$$

where N = M × M.

6. Time-Sharing-Based Synchronization for Color-Independent Visual-MIMO

An appropriate synchronization method for a color-independent visual-MIMO system was proposed previously [25]. Usually, LEDs send data at much higher speeds than the camera's frame rate. To solve this problem, we proposed an effective method of synchronization by time sharing information data with synchronization data, even though the camera frame rate is much lower than the desired data rate [25]. We were able to generate information and synchronization data as shown in Figure 10 without violating the purpose of GCM [25].

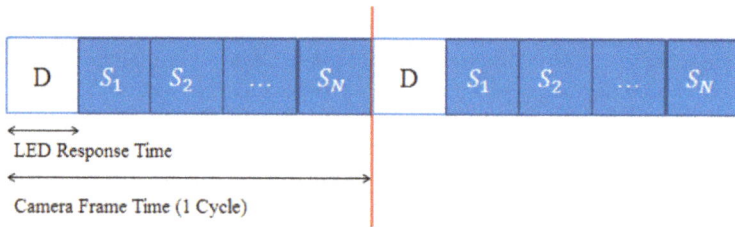

Figure 10. Structure of the transmitted data for time-sharing-based synchronization [25].

The white (D) and blue (S) sections represent information symbol (color) data and synchronization data, respectively. Here, $S_1 - S_N$ are symbols generated from the color-space-based constellation. Figure 11 presents the synchronization flow chart for a color independent visual-MIMO system on the receiver side [25].

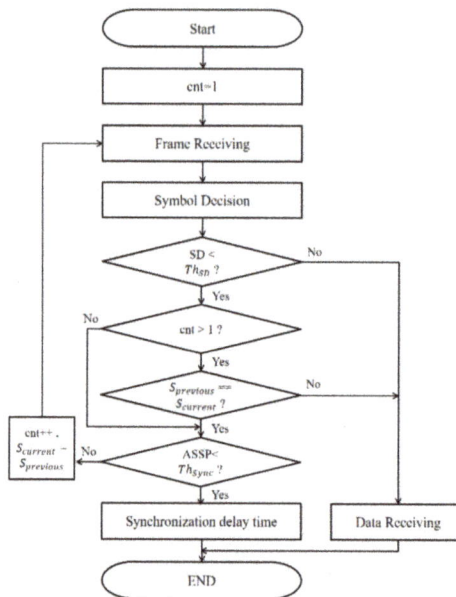

Figure 11. Synchronization flow chart for a color-independent visual-MIMO system on the receiver side [25].

This time-sharing-based synchronization can overcome synchronization and flickering problems in a color-independent visual-MIMO system. While maintaining the uniformity of color, synchronization is possible when the LEDs flicker at a higher rate than the camera's frame rate. This is an advantage for the human eye in that it is not able to detect this high-speed flickering. Therefore, this method not only maintains the function of the desired lighting source but also presents the possibility of communication with commercial cameras.

7. Case Studies of Color-Independent Visual-MIMO

Various applications are possible for visual-MIMO based communication. For example, novel advertising systems may use smartphone cameras to receive product information from electronic billboards. In addition, in museums, kiosks can display information for cell phone cameras, such as maps, images, and customized audio tours. This kind of communication system can also be used in vehicle-to-vehicle (or road), robot-to-robot, and hand-held displays to fixed surveillance cameras. In this section, we will discuss two applications of a color-independent visual-MIMO system: (1) fusion research combining fashion and technology and (2) V2X communication.

7.1. Wearable Visual MIMO

With the growth in the wearable device market, various wearable devices have utilized LEDs for wearable interaction. This is an attempt to combine color-space-based visual-MIMO systems with wearable devices in order to extend the use of LEDs in existing wearable devices to user-oriented interaction.

The design of LED light, which is a practical medium for the transmission of data in a VLC system, must be carefully considered because of its visible characteristics. Most of the existing VLC systems have been based on the use of white light, so the technology should not be simply applied to wearable devices (i.e., fashionable design) as it is. Communication using RGB LEDs, which are used in wearable devices, can be controlled by two parameters: lighting color and brightness. However, for designers, determining the color and brightness of an LED to be embedded in wearable smart fashion is a very important design issue. Thus, because the color and brightness of the light can vary in accordance with the fashion designer's wishes, seamless communication should be guaranteed for all variations of color and brightness. Given these factors, the color independency of GCM is an essential feature for fashion design. Designers can choose the color and brightness for their design, and engineers can use GCM to set the symbol color to match the chosen color and brightness. Figure 12 presents a transceiving block diagram of a GCM-based color-independent visual-MIMO system with the addition of an LED light design stage [26].

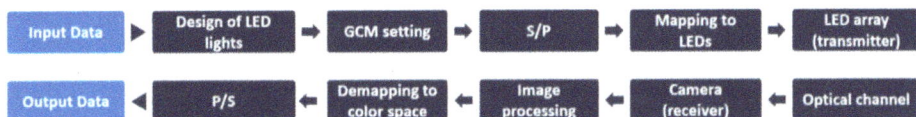

Figure 12. GCM-based visual MIMO transceiving process including the design of LED lights [26].

Figure 13 displays a fashion smart module using fasteners based on a color-independent visual-MIMO platform that is currently under development [26]. The LED array embedded in the strapped cuffs and the zipper slider acts as transmitter for the visual-MIMO system, and the data to be transmitted can be controlled using a smartphone connected with Bluetooth. The transmitted data can be received via a separate device (e.g., a smartphone) with a built-in camera.

The color-independent visual MIMO platform combined with wearable devices is expected to become another means by which users can express their personality in the customized and user-centered wearable device market. For this purpose, the systematic development of software algorithms and hardware design is required for the adaptive application of this technology to various wearable devices in clothing, shoes, and accessories.

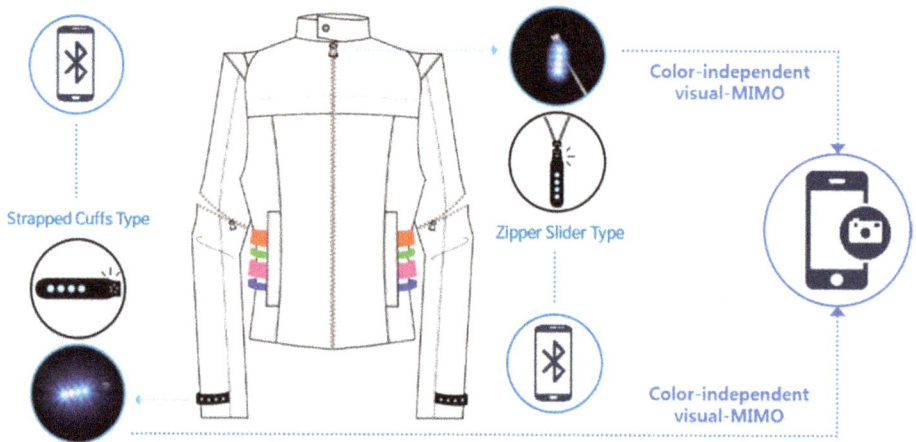

Figure 13. A smart module consisting of strapped cuff and zipper slider fasteners based on a color-independent visual-MIMO platform [26].

7.2. V2X Communication

The communication between a vehicle and other devices is another application of visual-MIMO systems. This approach has been described in our research work [23]. Figure 14 shows an example of V2X communication. We proposed two methods for V2X communication using visual-MIMO networking: multipath transmission and multinode (multiple access) communication. Multipath transmission can be used when the distance between the source and destination is far. Information transmitted to the destination node can be relayed via other nodes. For example, a relay vehicle can be used to transfer data. This feature of the network allows multipath transmission, similar to cooperative RF communication, but with lower coordination overhead [23]. We may also consider another multiple access communication in which a destination node can receive the information from multiple source nodes simultaneously. If the light sources of multiple nodes do not overlap within the field of view of the camera during V2X communication, it is possible to receive each light source by separating multiple sources through image processing techniques. To avoid the overlap of the multiple light sources, a divided region of interest (ROI) method by dividing communication area has been proposed [23]. Then, we may achieve the simultaneous multiple access communication without interchannel (inter-node) interferences.

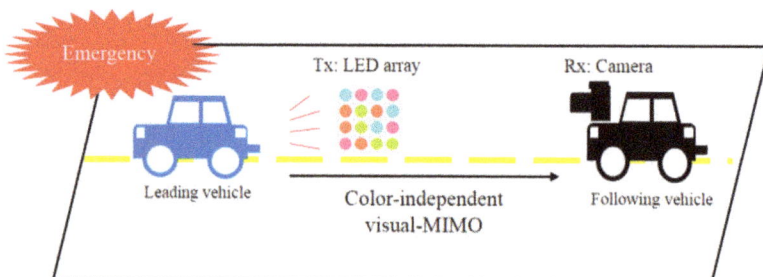

Figure 14. Example of V2X communication [23].

8. Future Research

LED color detection is a challenge in visual-MIMO systems. In a past paper [27], we used regression analysis in LED color detection for both high and low environmental light intensities, representing a simple machine-learning approach. In this strategy, we first detected each LED image from the whole LED array image. We then manually extracted some features for training the LED images to establish a multiple linear regression model. Finally, we obtained the multiple linear regression model to predict the color of the LED images. Using our trained multiple linear regression model, the color closeness accuracy was close to 90% for high and low environmental light intensities. For this work, we designed an experimental setup with a smartphone camera as a receiver, a light meter (CL-200A) for measuring environmental light intensity, and two tripods to fix the smartphone and transmitter in front of two light sources in a dark black room. The distance between the transmitter and smartphone was 30 cm. The details of the experimental setup and measurements are described in our previous work [27]. Based on this research [27], it is clear that machine learning is a promising approach for the visual-MIMO system. In the near future, this machine-learning approach can be used to predict LED color from LED images based on the distance and angle between the transmitter and receiver for different environmental light intensities. We also intend to measure and analyze the communication performance of this approach.

Visual-MIMO is the part of the VLC system. We have given some examples of visual-MIMO communication systems. Other researchers [28–30] focus on the sustainability of the indoor VLC in heterogeneous network. They pointed out technical challenges like handover, resource management, interference minimization, and optimization of channel capacity in this field. These topics are prominent to establish a complete network using VLC combining with an RF network. In the future, we will survey the suitability of our visual-MIMO scheme in indoor applications by considering above mentioned challenges.

9. Conclusions

In this paper, we first reviewed the design and performance of color-space-based modulation for a VLC system that is independent of the target color. This included a summary of current universal manipulation methods, such as color space selection, constellation generation, and mapping-demapping processes, that are uniquely suitable for the manipulation of light signals in the transceiving of data in typical VLC systems. We then described our research works which we underwent with the goal of achieving a color-independent visual-MIMO system that incorporates all of the advantages of GCM, generating higher data rates over larger distances, and improving performance by using image processing in addition to color independency. Besides, we described an algorithm for time-sharing-based synchronization that is vital for LED array-to-camera communication. We also demonstrated two practical applications of a GCM-based visual-MIMO system. Finally, we were able to establish that machine learning could be useful component of GCM based visual MIMO communication.

Author Contributions: The manuscript had been written by T.-H.K; he also designed the experiments. J.-E.K. helped to perform the experiments as well as analyze the results. Y.-H.K described the fusion research combining the color-space-based visual-MIMO systems with wearable devices. As the corresponding author, K.-D.K. proposed the idea as well as supervised the research; he wrote the manuscript.

Funding: This research was supported by the Basic Science Research Program through the National Research Foundation (NRF) of Korea funded by the Ministry of Education [2015R1D1A1A01061396] and was also supported by the National Research Foundation of Korea Grant funded by the Ministry of Science, ICT, Future Planning [2015R1A5A7037615].

Conflicts of Interest: The authors declare no conflicts of interest.

References

1. Komine, T.; Nakagawa, M. Fundamental Analysis for Visible-Light Communication System using LED Lights. *IEEE Trans. Consum. Electron.* **2004**, *50*, 100–107. [CrossRef]
2. Elgala, H.; Mesleh, R.; Haas, H. Indoor Optical Wireless Communication: Potential and State-of-the-Art. *IEEE Commun. Mag.* **2011**, *41*, 56–62. [CrossRef]

3. Ergul, O.; Dinc, E.; Akan, O.B. Communicate to illuminate: State-of-the-art and research challenges for visible light communications. *Phys. Commun.* **2015**, *17*, 72–85. [CrossRef]
4. Lee, K.; Park, H. Modulations for visible light communications with dimming control. *IEEE Photonics Technol. Lett.* **2011**, *23*, 1136–1138. [CrossRef]
5. Anand, J.M.; Mishra, P. A novel modulation scheme for visible light communication. In Proceedings of the 2010 Annual IEEE India Conference (INDICON), Kolkata, India, 17–19 December 2010; pp. 1–3. [CrossRef]
6. Hashemi, S.K.; Ghassemlooy, Z.; Chao, L.; Benhaddou, D. Orthogonal frequency division multiplexing for indoor optical wireless communications using visible light LEDs. In Proceedings of the 2008 6th International Symposium on Communication Systems, Networks and Digital Signal Processing (CSNDSP), Graz, Austria, 25 July 2008; pp. 174–178. [CrossRef]
7. Afgani, M.Z.; Haas, H.; Elgala, H.; Knipp, D. Visible light communication using OFDM. In Proceedings of the 2nd International Conference on Testbeds and Research Infrastructures for the Development of Networks and Communities (TRIDENTCOM 2006), Barcelona, Spain, 1–3 March 2006. [CrossRef]
8. Sugiyama, H.; Haruyama, S.; Nakagawa, M. Experimental investigation of modulation method for visible-light communication. *IEICE Trans. Commun.* **2006**, *E89-B*, 3393–3400. [CrossRef]
9. Khan, T.A.; Tahir, M.; Usman, A. Visible light communication using wavelength division multiplexing for smart spaces. In Proceedings of the 2012 IEEE Consumer Communications and Networking Conference (CCNC), Las Vegas, NV, USA, 14–17 January 2012; pp. 230–234. [CrossRef]
10. Yokoi, A.; Samsung Yokoham Research Institute. Color multiplex coding for VLC. In *IEEE P802.15 Working Group for Wireless Personal Area Networks (WPANs)*; Institute of Electrical and Electronics Engineers (IEEE): Piscataway, NJ, USA, 2008.
11. IEEE Standard for Local and Metropolitan Area Networks—Part 15.7: Short-Range Wireless Optical Communication Using Visible Light. *IEEE Stand. 802.15.7* **2011**. [CrossRef]
12. Drost, R.J.; Sadler, B.M. Constellation design for color-shift keying using billiards algorithms. In Proceedings of the 2010 IEEE GLOBECOM Workshops (GC Wkshps), Miami, FL, USA, 6–10 December 2010; pp. 980–984. [CrossRef]
13. Das, P.; Kim, B.-Y.; Park, Y.; Kim, K.-D. A New Color Space Based Constellation Diagram and Modulation Scheme for Color Independent VLC. *Adv. Electr. Comput. Eng.* **2012**, *12*, 11–18. [CrossRef]
14. Das, P.; Kim, B.-Y.; Park, Y.; Kim, K.-D. Color-independent VLC based on a color space without sending target color information. *Opt. Commun.* **2013**, *286*, 69–73. [CrossRef]
15. Das, P.; Park, Y.; Kim, K.-D. Performance improvement of color space based VLC modulation schemes under color and intensity variation. *Opt. Commun.* **2013**, *303*, 1–7. [CrossRef]
16. Das, P.; Park, Y.; Kim, K.-D. Performance analysis of color-independent visible light communication using a color-space-based constellation diagram and modulation scheme. *Wirel. Pers. Commun.* **2014**, *74*, 665–682. [CrossRef]
17. Ashok, A.; Gruteser, M.; Mandayam, N.B.; Silva, J.; Dana, K.; Varga, M. Challenge: Mobile optical networks through visual MIMO. In Proceedings of the MobiCom'10: Sixteenth Annual International Conference on Mobile Computing and Networking, Chicago, IL, USA, 20–24 September 2010; pp. 105–112.
18. Ashokz, A.; Gruteserz, M.; Mandayamz, N.; Dana, K. Characterizing Multiplexing and Diversity in Visual MIMO. In Proceedings of the 45th Annual Conference on Information Sciences and Systems (CISS), Baltimore, MD, USA, 23–25 March 2011; pp. 1–6. [CrossRef]
19. Berns, R.S. *Billmeyer and Saltzman's Principles of Color: Technology*, 3rd ed.; Wiley: New York, NY, USA, 2000; Chapter 2; ISBN 978-0-471-19459-0.
20. Elgala, H.; Mesleh, R.; Haas, H.; Pricope, B. OFDM visible light wireless communication based on white LEDs. In Proceedings of the 2007 IEEE 65th Vehicular Technology Conference—VTC2007-Spring, Dublin, Ireland, 22–25 April 2007; pp. 2185–2189. [CrossRef]
21. Lee, D.; Choi, K.; Kim, K.-D.; Park, Y. Visible light wireless communication based on predistorted OFDM. *Opt. Commun.* **2012**, *285*, 1767–1770. [CrossRef]
22. Das, P.; Park, Y.; Kim, K.-D. Performance of color-independent OFDM visible light communication based on color space. *Opt. Commun.* **2014**, *324*, 264–268. [CrossRef]
23. Kim, J.-E.; Kim, J.-W.; Park, Y.; Kim, K.-D. Color-Space-Based Visual-MIMO for V2X Communication. *Sensors* **2016**, *16*, E591. [CrossRef] [PubMed]

24. Kim, J.-E.; Kim, J.-W.; Kim, K.-D. LEA Detection and Tracking Method for Color-Independent Visual-MIMO. *Sensors* **2016**, *16*, 1027. [CrossRef] [PubMed]
25. Kwon, T.-H.; Kim, J.-E.; Kim, K.-D. Time-Sharing-Based Synchronization and Performance Evaluation of Color-Independent Visual-MIMO Communication. *Sensors* **2018**, *18*, E1553. [CrossRef] [PubMed]
26. Kim, J.-E.; Kim, Y.-H.; Oh, J.-H.; Kim, K.-D. Interactive Smart Fashion using User-oriented Visible-Light Communication: The Case of Modular Strapped Cuffs and Zipper Slider Types. *Wirel. Commun. Mob. Comput.* **2017**, *2017*, 1–13. [CrossRef]
27. Banik, P.P.; Saha, R.; Kim, K.-D. Regression analysis for LED color detection of visual-MIMO system. *Opt. Commun.* **2018**, *413*, 121–130. [CrossRef]
28. Seguel, F.; Soto, I.; Iturralde, D.; Adasme, P.; Nuñez, B. Enhancement of the QoS in an OFDMA/VLC system. In Proceedings of the 2016 10th International Symposium on Communication Systems, Networks and Digital Signal Processing (CSNDSP), Prague, Czech Republic, 20–22 July 2016; pp. 1–5. [CrossRef]
29. Tsiropoulou, E.E.; Vamvakas, P.; Papavassiliou, S. *Resource Allocation in Next-Generation Broadband Wireless Access Networks*; Singhal, C., De, S., Eds.; IGI Global: Hershey, PA, USA, 2017; Chapter 10.
30. Lin, B.; Tang, X.; Ghassemlooy, Z.; Lin, C.; Li, Y. Experimental Demonstration of an Indoor VLC Positioning System Based on OFDMA. *IEEE Photonics J.* **2017**, *9*, 1–9. [CrossRef]

electronics

Article

Visible Light Communication System Based on Software Defined Radio: Performance Study of Intelligent Transportation and Indoor Applications

Radek Martinek *,†, **Lukas Danys** *,† and and **Rene Jaros** *,†

Department of Cybernetics and Biomedical Engineering, Faculty of Electrical Engineering and Computer Science, VSB–Technical University of Ostrava, 17. listopadu 15, 708 33 Ostrava, Czech Republic
* Correspondence: radek.martinek@vsb.cz (R.M.); lukas.danys@vsb.cz (L.D.); rene.jaros@vsb.cz (R.J.);
  Tel.: +420-721-009-971 (R.M.); +420-734-239-361 (L.D.); +420-774-650-522 (R.J.)
† These authors contributed equally to this work.

Received: 28 February 2019; Accepted: 10 April 2019; Published: 15 April 2019

Abstract: In this paper, our first attempt at visible light communication system, based on software defined radio (SDR) and implemented in LabVIEW is introduced. This paper mainly focuses on two most commonly used types of LED lights, ceiling lights and LED car lamps/tail-lights. The primary focus of this study is to determine the basic parameters of real implementation of visible light communication (VLC) system, such as transmit speed, communication errors (bit-error ratio, error vector magnitude, energy per bit to noise power spectral density ratio) and highest reachable distance. This work focuses on testing various multistate quadrature amplitude modulation (M-QAM). We have used Skoda Octavia III tail-light and Phillips indoor ceiling light as transmitters and SI PIN Thorlabs photodetector as receiver. Testing method for each light was different. When testing ceiling light, we have focused on reachable distance for each M-QAM variant. On the other side, Octavia tail-light was tested in variable nature conditions (such as thermal turbulence, rain, fog) simulated in special testing box. This work will present our solution, measured parameters and possible weak spots, which will be adjusted in the future.

Keywords: multistate quadrature amplitude modulation (M-QAM); visible light communication (VLC); software defined radio (SDR); sofware defined optics (SDO); LED tail-light; LED indoor ceiling light; vehicle-to-everything (V2X); nature conditions (thermal turbulence, rain, fog); bit-error ratio (BER)

---

1. Introduction

In recent years, visible light communication (VLC) surfaced as an alternative to classical radio frequency (RF) technology [1–3]. Current communication bands often lack free channels, which is notable particularly in Wi-Fi or in industrial, scientific and medical bands. VLC is an optical wireless standard which operates from 380 to 780 nm, using a visible light source as a signal transmitter, free space environment as transmission medium and the appropriate photodiode/photodetector as a receiver. VLC seems to be capable technology for short-range or possibly in the future even long-range communications. Future appliances vary greatly, spanning from vehicle-to-vehicle [4–9] communications, infrastructure-to-vehicle communications or simply as an alternative to typical local area networks (LAN) [10,11].

A number of papers focused on multiple VLC technologies [12]. Light-fidelity (Li-Fi) [13–20] is slowly surfacing as commercially available alternative to Wi-Fi [21]. Orthogonal frequency division multiplexing in car-to-car was tested in real-world driving scenarios by Shen et al. [22].

VLC on software defined radio (SDR) [23,24] is evolving quickly. In 2011, a 1 Mbps video stream was achievable over 3 meters, when deployed on custom LED matrixes [25]. In 2015, Hussain et al.

tested the implementation of IEEE 802.15.7, they achieved results according to this standard, however transmission distance was limited to 1 m [26]. Nowadays, we are testing longer distances and mainly higher data rates, even on commercial light sources.

Rapid expansion of LED is crucial for this technology, as it offers multiple advantages such as long lifespan, low power consumption, high tolerance to humidity, high efficiency, and fast switching. However the main advantage of VLC based on LED is the use of the visible spectrum (380–780 nm). For this reason LED can perform communication functionality while maintaining the original function as illumination lighting. In this paper, we used commercially available and currently used light sources. For this purpose, a Skoda Octavia III tail-light without any modifications was chosen. To test indoor deployment, we have also used Phillips Fortimo DLM 300 44 W/840 Gen3 [27]. Avalanche photodiodes (APD) and positive-intrinsic-negative (PIN) detectors are commonly used as receivers. We used Thorlabs PDA36A-EC PIN [28] photodetector with 13 mm^2 of active area as receiver, since it was the most suitable candidate from available portfolio.

Our work is aimed at implementation of a vehicle-to-everything (V2X) system with highly modular design [29–31]. For this reason, we have developed a system, based on SDR. Each individual component can be swiftly exchanged, without any necessary adjustments to original code. Concept of V2X is based on the passing of information from a vehicle to any appropriate entity and vice versa. It is also often divided into different subsections, such as vehicle-to-infrastructure (V2I) [32,33], vehicle-to-network (V2N) [34,35], vehicle-to-vehicle (V2V) [36,37], vehicle-to-pedestrian (V2P) [35,38], vehicle-to-device (V2D) [39,40] and vehicle-to-grid (V2G) [41,42], vehicle-to-home (V2H) [43,44]. To test these concepts, we have also built our own testing polygon named BroadBAND light. As Skoda cars are the most widespread vehicles in the Czech Republic, we have picked Skoda Octavia III tail-light as the transmitter in V2X scenarios. So far, the partnership with manufacturers of these lights have yielded results, as we have received multiple samples of planned or already available products for testing. Philips Fortimo DLM300 is the most deployed LED indoor ceiling light in Czech Republic, so we used it in indoor experiments. Most Czech public institutions deploy exactly this type–testing it is the most logical step, as we can use every ceiling light at our university as transmitters.

Figure 1 describes different ways of V2X communication [30,45,46], mentioned earlier. V2V is a system, which enables car to communicate with each other. Its main goal is to reduce vehicle collisions and crashes. It will be a backbone of multiple levels of autonomy, delivering assisted driver services like collision warnings [32,47,48]. One key issue with V2V is that to be most effective, it should reside in all cars on the road. However, this technology has to start somewhere, so car manufacturers are slowly introducing their solutions. V2D is a system, that links cars to many external receiving devices and will be particularly useful to bikers. Vehicles can communicate with V2D device on cycle to alert rider to potential danger or to avoid accidents [39,40,49]. V2P is a system which should be particularly useful to elderly persons, school kids and physically challenged persons. V2P maintains a link between pedestrian's smart devices and vehicles to act as an advisory to avoid collisions [50,51]. V2H communications involve a link between a vehicle and the owner's home, sharing the task of providing energy [52,53]. During power outages, a vehicles battery can be used as a critical power source. V2G is a system which can communicate with the power grid to adjust the vehicle's charging rate [54,55]. It will be an element in some electric vehicles and is used as a power grid modulator to dynamically adjust to the energy demand [56].

Figure 1. Vehicle-to-everything scenarios communication.

Figure 2 describes proposed different ways of indoor visible light communication (VLC) communication. Smart lighting inside smart buildings provides the infrastructure for illumination, control and communications and will greatly reduce energy consumption within a building. Smart appliances, meters or factory applications, especially in dangerous conditions are all possible target devices. Direct connection between mobile devices might be possible as well, since modern smartphones have illumination LEDs which might be used as transmitters and cameras, which could work as receivers. There are advantages of using VLC in hospitals and healthcare [57–60]. RF technologies are mostly undesirable in certain parts of hospitals, especially around MRI scanners and in operating theatres. VLC deployment in aviation is also desirable. Radio is undesirable in passenger compartments of aircrafts. Modern aircrafts already use LEDs for illumination, so VLC might replace wiring to passenger seats. This might reduce aircraft construction costs and weight [61,62].

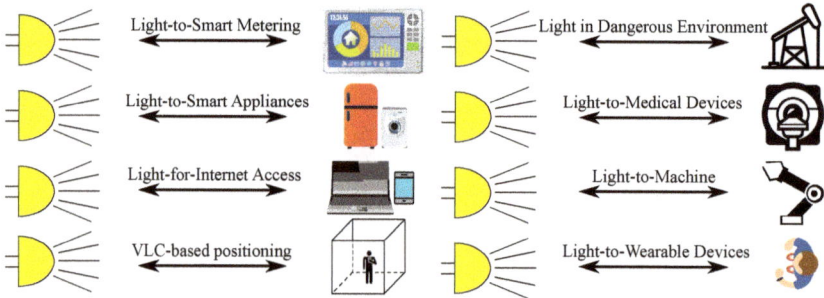

Figure 2. Indoor VLC scenarios.

Factory applications are especially interesting. There are multiple benefits to machine to machine VLC solutions (machine-to-machine communication - M2M [20,63–66]). Smart factories could be built upon enhanced industrial instrumentation, such as advanced sensors or meters. VLC can be also used to coordinate the movement and timing of robots in manufacturing settings, such as automobile factories. It can provide location-based communications to automated guided carts and "smart cart" robots or provide drone-to-station communications for precise, interference-free movement, drop zones and landing sites [67,68].

Vehicular visible light communication (V-VLC) [46,69–73] is often described as supplementary or "sister" technology to planned 5G [31]. There is also a concept of social internet of vehicles (SIOV) [74–76], which is based on LTE/4G network, consisted of multiple road side units (RSU) based on eNodeB base stations and on-board units (OBU) with LTE/4G capabilities. SIOV consists of multiple entities, which are treated as nodes connected to each other via links. However this concept can be further expanded by VLC implementation. Daytime running lamps (DRL) are often mandatory in European countries and optional VLC capabilities would possibly save energy needed to maintain continual 4G/5G connection [77,78].

Multiple teams implemented VLC using SDR and LabVIEW, with different success. A turkish team successfully carried out a number of experiments, using simpler on-off keying (OOK) and variable pulse position modulation (VPPM) modulations [79]. Another team from Chile managed to implement their own system using similar hardware [7]. However, there are two main concerns. Both teams do not mention their maximal reached transmit speed and the used photodetector (PDA36A) is certainly a limiting factor, as we have run a number of tests on it as well. According to manufacturer's datasheet, increasing gain of its trans-impedance amplifier significantly limits useable bandwidth. According to the state of art study, universal software radio peripherals(USRPs) are more than capable of being used in VLC experiments [80]. They also mention the necessary adjustments and possible limitations of lower end (USRP 292X) models. That's why we switched to much more powerful and newer models which will be used in future experiments [7,79–82].

As mentioned before, the described system is our team's first attempt at SDR VLC quadrature amplitude modulation(QAM) system. Different teams have already delved into different problematics, such as adaptive modulation schemes [83]. Multiple input multiple output design described by Deng et al. 2017 [84] is already implementing higher state QAM modulation schemes capable of adaptive switching. Designing similar systems based on orthogonal frequency division multiplexing (OFDM) will be a topic of imminent research, as we are already switching to more capable hardware components [85,86]. Khalid et al. investigated the implementation of a VLC system based on DAQ hardware [87]. This system is however very limited performance-wise. The second iteration of our prototype will be also based on LabVIEW, but we are focusing on field programmable gate array (FPGA) implementation, as it will increase performance dramatically.

There are multiple advantages to practical implementation of VLC. Each of presented solution has its own use and advantages. When approaching the problems of V2V communication, the natural conditions are major concerning factor, as they significantly vary throughout the year. Carrying out a number of experiments in this area is a logical first step in implementation of channel equalization, as we estimate it will significantly improve transmit speeds or reachable distance. Also we tested our modular platform in previously mentioned conditions, as it will become a basic platform, which will be modified and improved in the future. Currently, we are exploring outdoor car lamps, tail-lights, and indoor ceiling lights, as these light sources that are the most perspective. In the future we will also include street lamps, as V2I and V2V outdoor experiments will be carried out in previously mentioned testing polygon, which is already running on LEDs, so every lamp is prepared for VLC.

Tsiropoulou et al. [88] have investigated problematics of non-orthogonal multiple acces (NOMA) vs. orthogonal frequency division multiple access (OFDMA) approach [89–91]. According to her, NOMA offers us multiple advantages, such as considerable interference mitigation or simultaneous bandwidth utilization. Also due to the absence of resource block per user, NOMA can sufficiently accommodate more users than OFDMA. Modern mobile networks based on LTE use the OFDMA approach, as it is a basis of LTE standardization. Car manufacturers are currently enrolling LTE modules into vehicles and are preparing for 5G transfer. Currently, 5G is surfacing technology, which was deployed in only a limited number of countries. However 5G is based on NOMA, so its implementation in VLC is necessary if both technologies should coexist or work in conjunction. Lin et al. also tested hybrid NOMA/OFDMA approach with partial success [92].

Tsiropoulou et al. [93] propose a concept of visible light communication local area networks (VLC-LANs), where users are served by optical access points (OAPs). In this scenario, VLC-LANs [94] are presented as alternative to macrocell area coverage. Two-tier VLC topology is considered, mostly for indoor and outdoor coverage. The system is based on OFDMA, basically to be ready for LTE incorporation. Each user communicates directly with a single OAP via communication link. OAPs total bandwidth is divided into subcarriers, which are organized in resource block (RB). Each RB is occupied exclusively by one user. This concept could be easily adapted for 5G by introduction of NOMA, as discussed earlier.

2. Experimental Setup

Experimental setup is based upon application created in LabVIEW, which was used for input/output signal processing and measurement. Application output was fed into NI USRP-2921 [95], which stands for transmitting element. Signal was then amplified by a 1.6 W amplifier, effectively working from 1 to 200 MHz. The amplified signal went through bias tee into the transmitting light source, which can be exchanged at will. A photodetector was located at a variable distance from the transmitting element. It was also possible to insert special measuring box, used for simulation of multiple nature conditions, such as fog, rain or thermal turbulence. Received signal was fed into NI USRP-2921 whose output is connected to the same computer running LabVIEW application. Signal was then evaluated, and parameters were displayed accordingly. This whole setup was designed with the highest possible modularity in mind. We wanted to switch each component at will and observe their direct impact on whole prototype.

We had to swap NI USRP-2921 transmitting and receiving boards with Ettus LFRX/LFTX Daughterboards, which were operating at 0-30 MHz. This band was approximately what we had in mind when designing whole prototype, as USRP capabilities were one of the biggest limiting factors. Original boards from 2921 were designed for 2.4–2.5 GHz and 4.9–5.9 GHz, which was completely unsuitable for our needs. We had also adjusted connectors on the Octavia tail-light, as they were designed for car engine control unit. To further improve RSL when transmitting using tail-light, we had mounted a planoconvex lens on a photodetector to focus received light into the converging beam with lens focus at the active area of photodiode. However, since we measured RSL at variable receiving angle when using stationary ceiling light as source, it was undesirable to use lens in this different scenario. We have chosen ZX85-12G+ bias tee [96], which operates from 0.2 MHz to 12 GHz, its maximal current is 0.4 A and maximal input voltage is 25 V. There are multiple parameters, that can be configured in LabVIEW application, such as:

- Carrier frequency: max 30 MHz
- Bandwidth
- Sample width
- Number of states: max 4096-QAM
- Message symbols
- Used TX filter
- TX gain
- RX gain
- TX device IP address
- RX device IP address

Tests wre carried out using static modulation formats. Long term measurements were essential to specify threshold for successful modulation switching in adaptive modulation. We were aiming for a similar system, which is used in case of microwave point-to-point links, where both units are capable of quickly changing modulation scheme according to natural conditions and measured parameters. Sacrificing part of the transmission speed in favor of link robustness is the main concept of this system.

Evaluation Parameters

Received signal level (RSL), which signalize signal strength received at second, or receiver, USRP. It is the sum of all losses and gains on the receiver input.

Our LabVIEW application use channel coding with hard decision forward error correction (FEC) threshold of 3.8×10^{-3}. Data with bit-error ratio (BER) below this threshold can be repaired by FEC codes. This threshold was appropriate for our M-QAM modulation. LabVIEW Modulation Tookit [97] also includes multiple different types of channel coding [98]. The influence of these techniques on transmission quality will be a topic of further research. As FEC was not the main topic of this paper, we used predetermined threshold and functions already present in LabVIEW libraries. Problematics of FEC in QAM VLC was investigated by team in Edinburg [99].

Modulation error ratio (MER) [100] is defined as a relationship between error vector magnitude (EVM) and signal-to-noise ratio (SNR). It was used to quantify the performance of transmitter/receiver in the system, which used digital modulations (in this case QAM). It is influenced by various imperfections in the implementation (such as noise, phase noise, distortion, right focusing) and characteristics of signal path which cause the actual constellation point to deviate from ideal locations.

EVM [101] is a measurement of demodulator performance in the presence of impairments (Figure 3, where $\vec{v}$ is the ideal symbol vector, $\vec{w}$ is the measured symbol vector, $\vec{w}-\vec{v}$ is the magnitude error, Φ is the phase error, and $\vec{e} = \vec{w} - \vec{v}$ is the error vector). The soft symbol decisions obtained after decimating the recovered waveform at the demodulator output were compared against ideal symbol locations. The root mean square error vector magnitude and phase error were then used in determining the EVM measurement over a window of N demodulated symbols. EVM was related to the modulation error ratio. There is one-to-one relationship between EVM and MER.

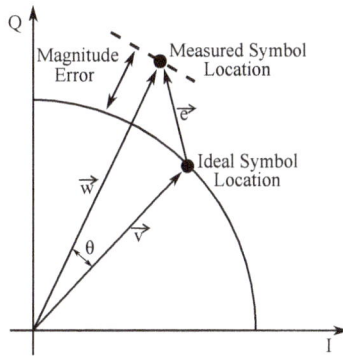

Figure 3. Error vector magnitude.

Energy per bit to noise power spectral density ratio (E_b/N_0) was an important parameter in digital communications. It is a normalized SNR measure, also known as SNR per bit. It is useful when comparing BER performance of different digital modulation schemes without taking bandwidth into account. To calculate this value, SNR must be periodically obtained at receiver USRP. E_b/N_0 was calculated at the receiver from the estimation of the SNR, the relationship between them is the following:

$$\frac{E_b}{N_0}\Big|_{dB} = SNR|_{dB} + 10log_{10}(n) - 10log_{10}(sps), \tag{1}$$

where n is the amount of information bits per symbol. Both modulation order and the code rate (in case that channel coding is employed) have an influence on this value. E_b/N_0 measurement on USRP is further expanded in paper by Alonso et al. [102].

3. A Feasibility Study on Indoor Visible Light Communication

Measurements using ceiling light were carried out under laboratory conditions (calm wind, 24 °C). Figure 4 describes tested setup. Our tested light source is capable to cover a conical area with radius of approximately 350 cm, as is described on Figure 5. We began our measurements at a right center of this covered area, directly under light source. Then, we periodically repeated measurements but moved to the edge of covered area with a step of 25 cm. The distance between receiver and transmitter was 202 cm.

The following configuration was used in this scenario:

- Carrier frequency: 3 MHz
- Bandwidth: 1–4 MHz
- Modulation type: M-QAM
- TX/RX gain: 0 dB
- Message symbols: 10,000
- Tx filter: root raised cosine
- Sample width: 16-bit
- Receiver-transmitter distance: 202 cm
- Measured distance: 0–350 cm (step of 25 cm)
- Photodetector without planoconvex lens → more suitable for this scenario
- Measured parameters: E_b/N_0, BER, EVM and MER

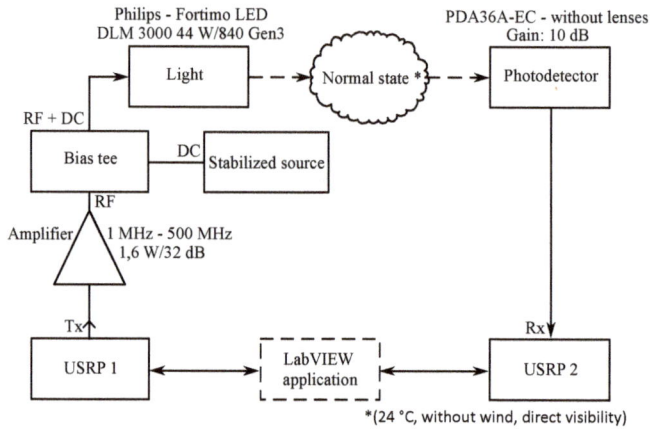

Figure 4. Ceiling light setup.

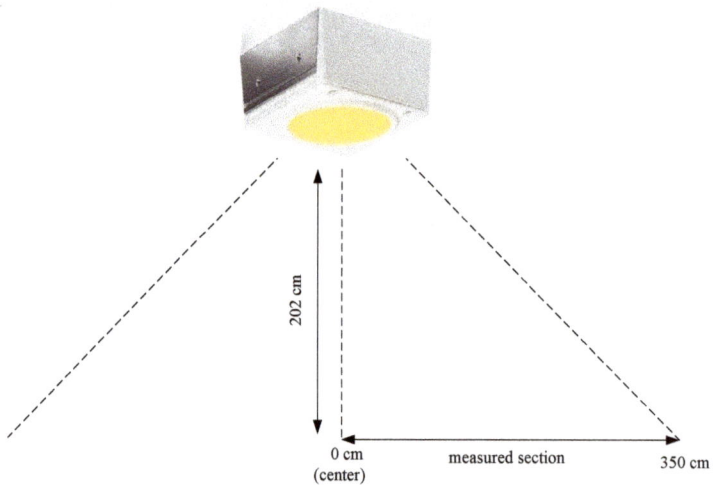

Figure 5. Setup with measured distance.

In Table 1 you can see parameters of light of Fortimo LED DLM 3000 44 W/840 Gen3 and in Table 2 you can see chosen parameters of photodetector PDA36A-EC.

Table 1. Parameters of light of Fortimo LED DLM 3000 44 W/840 Gen3.

Initial lumens	3000 lm
Color rendering index	80
Correlated color temperature	4000 K
System input power	46.0 W
System efficiency	68.0 Lm/W
Input voltage	220–240 V
Max. vitality	5000 h
U_{max} dc	80 V
Max operating temperature	65 °C
P_{max}	40 W

Table 2. Chosen parameters of photodetector PDA36A-EC.

Type of detector	Si PIN
Active area	3.6×3.6 mm (13 mm^2)
Wavelength range	350–1100 nm
Gain adjustment range	70 dB
Gain step	8×10 dB
Operating temperature	0–40 °C
Max output current	100 mA

Before proceeding with the main part, we measured attenuation characteristics of our prototype by using the vector network analyzer. Figure 6 represents measured data. It is visible that attenuation quickly increased with longer distances. Even the sample with the best conditions (directly under light source) showed an increase in attenuation by 47 dB relative to the reference of 0 dB. By increasing distance to 3 m, attenuation reached 68 dB, which negatively impacted prototype capabilities, mainly the constellation decoding. We have chosen carrier frequency of 3 MHz, so that we can increase bandwidth up to 4 MHz without getting into sub 1 MHz band. However, increasing bandwidth also increases the difference in attenuation between highest and lowest frequency. This fact negatively impacted constellation decoding as well. The resulting theoretical transmit speed was influenced by many factors, such as modulation scheme, used bandwidth and mainly by the distance of photodetector from center of measurements, which affected attenuation. We were able to reach 2 Mbps at 325 cm, by using 4-QAM with bandwidth of 1 MHz. The highest achieved transmit speed was 20 Mbps (32-QAM, 4 MHz), which could be maintained up to 90 cm from the center. Figure 7 represent possible maximal reachable transmission speeds for different M-QAM and bandwidths. Further implementation of adaptive modulation seems like a good way to further push this prototype forward. By defining strict rules for modulation switching, we could seamlessly maintain the highest possible transmit speed while having reliable connection. For our testing, we have used BER = 10^{-5} as a threshold value.

The valubble in Figure 7 are theoretically achievable transmit speeds. They were calculated in LabVIEW software, but a number of measurements were carried out to verify them. Effective transmit speed varied a bit (by approximately hundreds of kbps), because it was practically impossible to design an ideal channel. However deviations were so small, it was possible to neglect them.

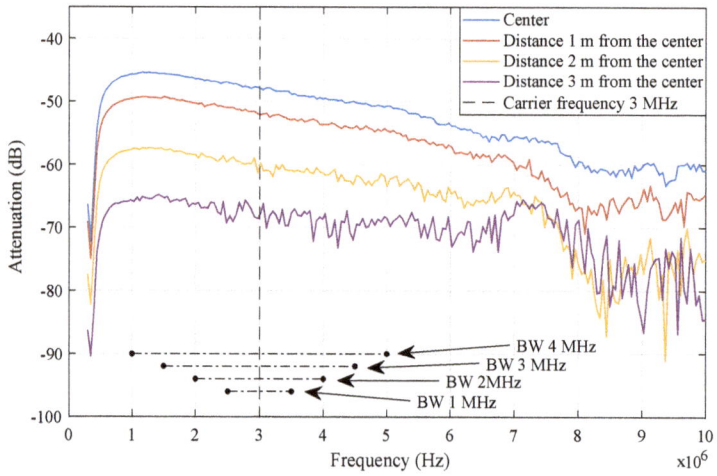

Figure 6. Attenuation characteristics of ceiling setup.

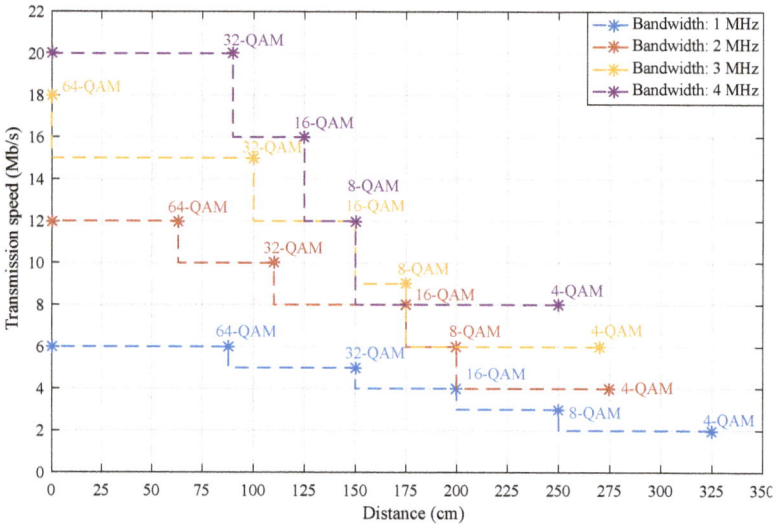

Figure 7. Transmit speed/distance from center relationship for ceiling light with different M-QAM and bandwidths.

Figure 8 represents a constellation diagrams of 4-QAM modulation measured at different distances and channel widths. It is noticeable that constellation points of the 4 MHz channel width measurements were more spread from their ideal position. These diagrams did not exceed the BER threshold mentioned earlier, but implementation of higher state modulation at limiting distance would result in rapid increase of BER and might even end in complete link degradation.

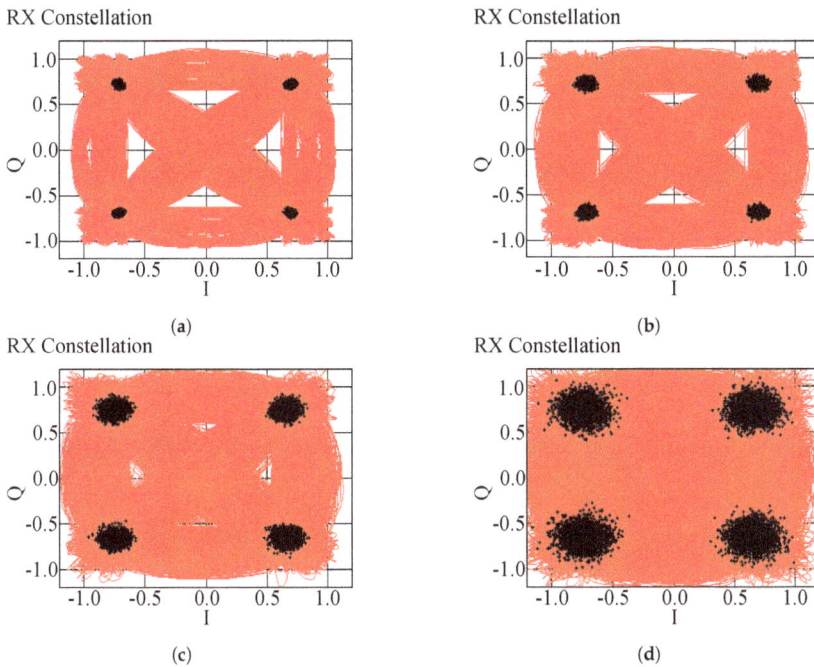

Figure 8. 4-QAM constellation diagrams of ceiling light and different bandwidths: (**a**) 4-QAM, BW = 1 MHz, (**b**) 4-QAM, BW = 2 MHz, (**c**) 4-QAM, BW = 3 MHz, (**d**) 4-QAM, BW = 4 MHz.

Figure 9 describes BER values of multiple channel widths and M-QAM modulation combinations. The graph itself also shows highest possible communication distance for measured modulation, which did not exceed BER threshold. The red line represents the FEC limit, so values exceeding it are beyond capabilities of fast error correction algorithms. Values located next to the arrows are BER.

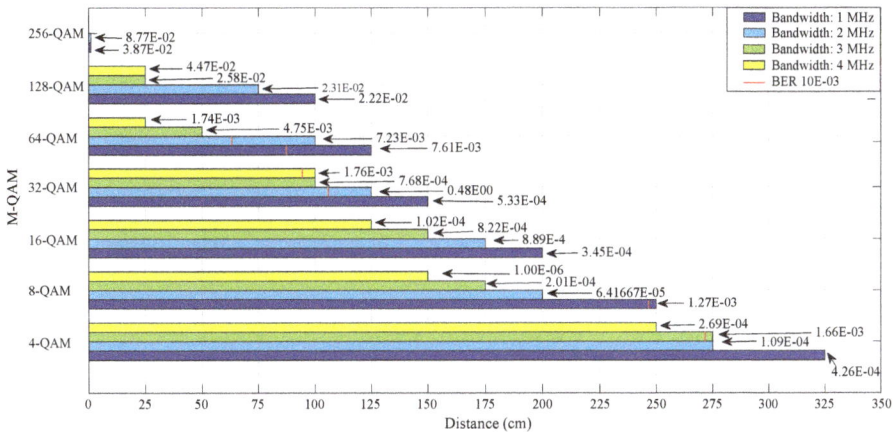

Figure 9. BER/distance from center relationship for ceiling light with different M-QAM and bandwidths.

Figures 10 and 11 represent a comparison of 1 MHz and 4 MHz channel widths, the highest and lowest measured variants. It is visible that E_b/N_0 values slowly decreased with increasing measured distance. In comparison, EVM values shows opposite trend. Wider channels were also much more limited in maximal reachable operation distance, which is also visible on Figure 7 as well. Simpler modulation schemes in combination with narrow channel are much more robust, which is visible on Figure 9.

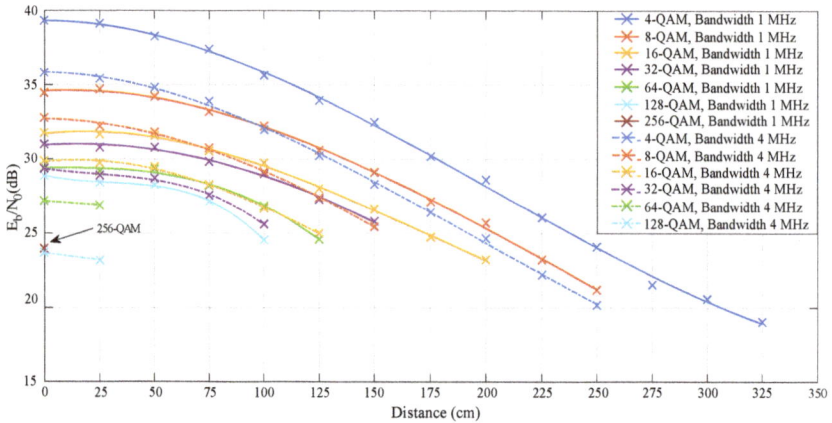

Figure 10. E_b/N_0 / distance from center relationship for ceiling light with different M-QAM and bandwidths.

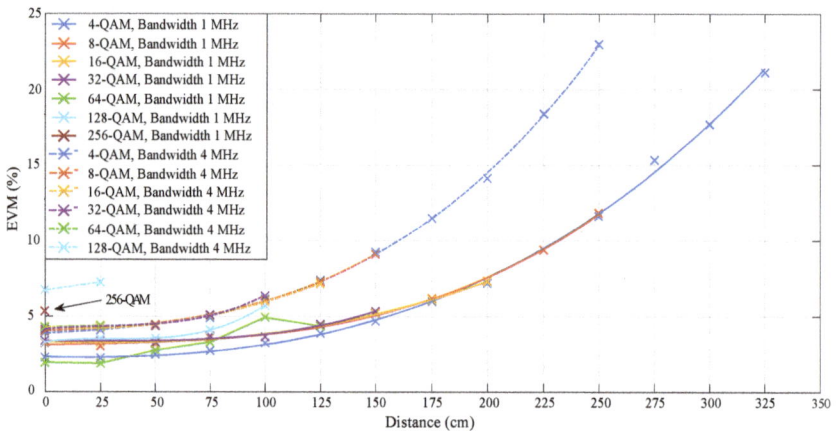

Figure 11. EVM/distance from center relationship for ceiling light with different M-QAM and bandwidths.

4. A Feasibility Study on Outdoor Visible Light Communication: Car Tail-Light

Measurements using car tail-light were carried out in multiple meteorological conditions. We approached this problem differently, as in real-life scenarios, meteorological conditions tend to often vary. These conditions were simulated in a plastic box made of plexiglass with dimensions of $50 \times 50 \times 500$ cm. Maximal measured distance in this scenario was 550 cm, which was effective threshold for functional communication. The first step was to measure the empty box as reference values for future comparisons. After that, three scenarios were simulated: fog, thermal turbulence and rain. These scenarios will be discussed further in this paper.

Figure 12 describes the tested setup, modified for Octavia tail-light. The whole experimental setup was very similar to the one used in ceiling light measurements. However, to carry out different scenarios the simulation box was inserted between transmitting light and receiving photodetector. Also using plano-convex lens was suitable for reaching higher RSL values and thus better transmit speed.

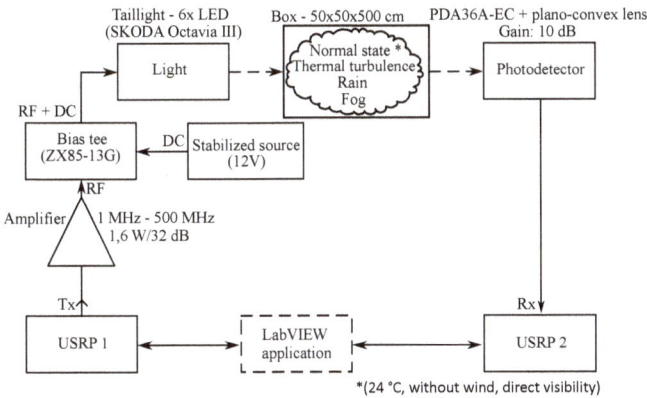

Figure 12. Octavia tail-light setup.

The following configuration was used in this scenario:

- Carrier frequency: 3 MHz
- Bandwidth: 1–4 MHz
- Modulation type: M-QAM
- TX/RX gain: 0 dB
- Message symbols: 10,000
- Tx filter: root raised cosine
- Sample width: 16-bit
- Receiver-transmitter distance: 550 cm
- Measured distance: 0–350 cm (step of 25 cm)
- Photodetector with planoconvex lens → more suitable for this scenario
- Measured parameters: E_b/N_0, BER, EVM, MER

Table 3 shows parameters of LED - LA G6SP. A matrix of these LEDs is used in Octavia tail-lights.

Table 3. Parameters of LED - LA G6SP.

Technology	InGaAlP Thinfilm
Viewing angle at 50 % IV	120 (Lambertian Emitter)
Color	Red (623 nm)
Optical efficiency	56 lm/W
Operating temperature	from −40 to 110 °C
Max surge current	1000 mA
Max current	140 mA
Typical voltage	2.1 V

Each measurement consisted of attenuation characteristics, which were measured by a vector network analyzer. By comparing every scenario, it was visible that fog had the highest impact, followed by rain and thermal turbulence, as seen on Figure 13. Also, we have included a reference power level of laser needed for functional 4-QAM transmission in the fog scenario.

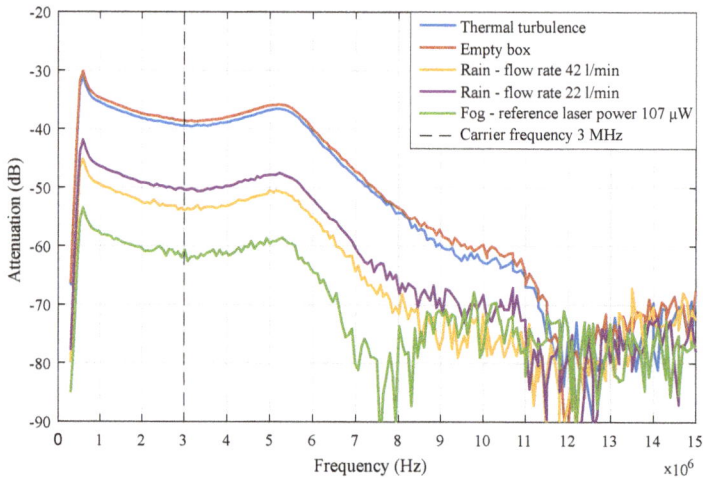

Figure 13. Attenuation characteristics of the Octavia tail-light.

Following measurements should cover most situations, which might happen in infrastructure-to-vehicle (I2V), V2V, V2I , V2N, V2P, V2D, V2G and V2H scenarios.

4.1. Scenario 1-Empty Box

As mentioned earlier, these measurements were used as reference values for further comparisons. To ensure conformity of measured data, these conditions were set as referential: 24 °C, direct visibility, windless. Setup can be seen in Figure 14.

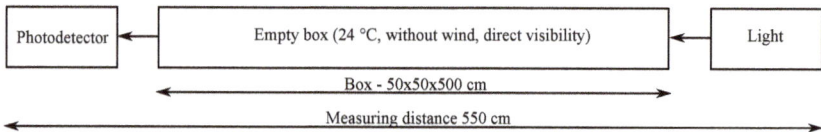

Figure 14. Octavia taillight setup adjusted for scenario 1—empty box.

Comparison of BER values with different bandwidths can be seen on Figure 15. It is noticeable that up to 64-QAM, BER values tended to stay below 10^{-5} threshold set earlier. On the contrary, 128-QAM and higher modulations suffered from much higher BER values. Bit error ratio of 256-QAM was even below 10^{-2} in every bandwidth combination. These values were too high for FEC and communication is impossible in these conditions. Figures 16 and 17 describe E_b/N_0 and EVM values for every M-QAM and bandwidth combination. It is noticeable that starting from 256-QAM modulation, the difference between each bandwidth tended to increase rapidly.

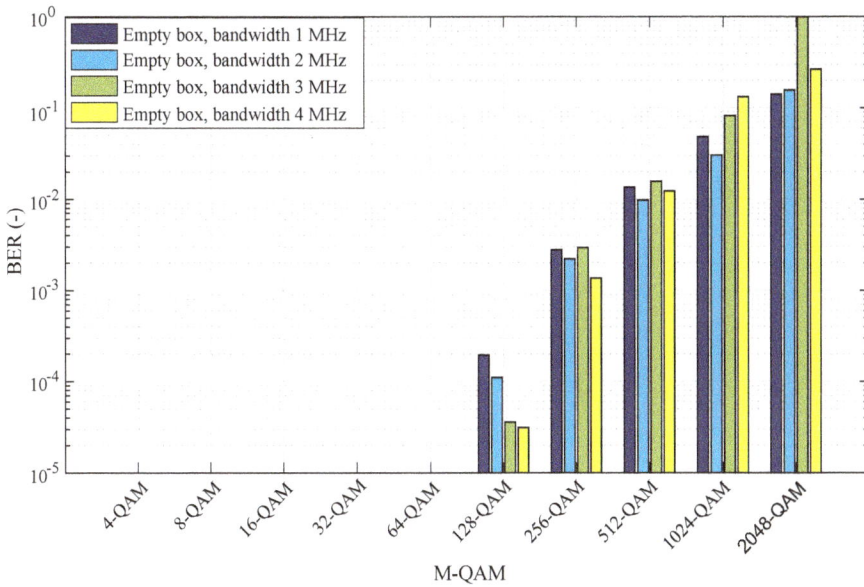

Figure 15. BER/distance relationship for Octavia tail-light with different M-QAM and bandwidths —scenario 1—empty box.

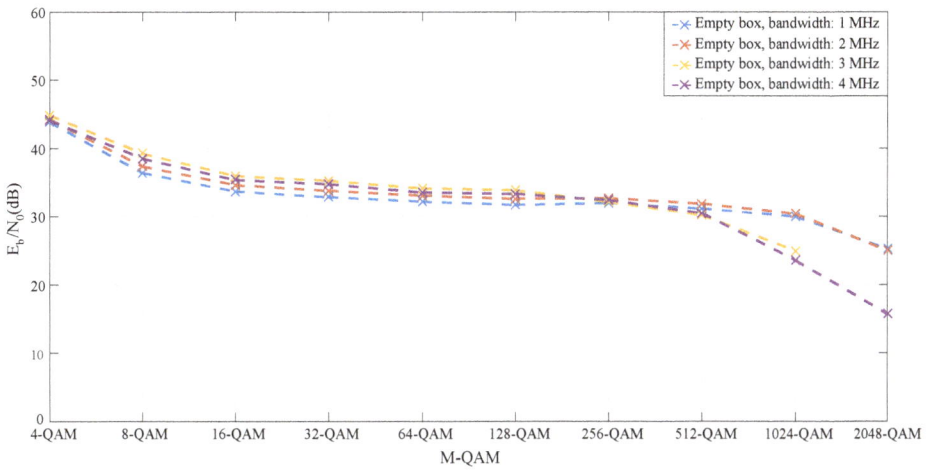

Figure 16. E_b/N_0 / distance relationship for Octavia tail-light with different M-QAM and bandwidths —in scenario 1—empty box.

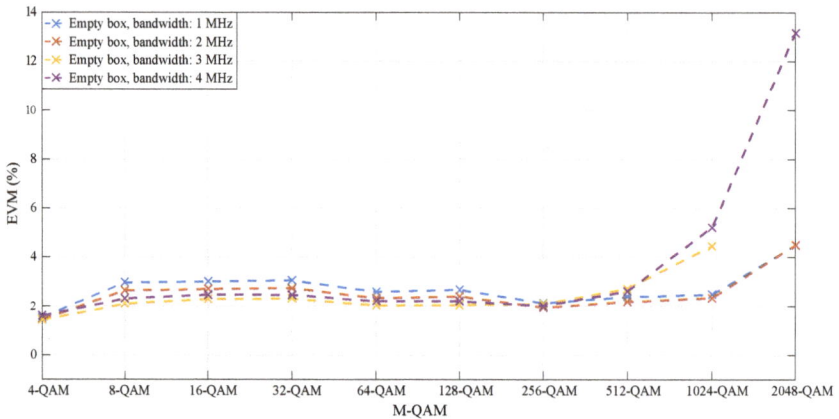

Figure 17. EVM/distance relationship for Octavia tail-light with different M-QAM and bandwidths—in scenario 1—empty box.

4.2. Scenario 2-Thermal Turbulence

Measurements with thermal turbulence were carried out with modified box. The bottom part was removed, and multiple hot-air blowers were mounted instead. The top part of box was perforated to ensure sufficient air flow. Each blower heated the air to 50 °C. Heated air then steadily flowed through box and slowly cooled to 44 °C, which were measured directly at perforations. The horizontal airflow was 0.3 m/s and vertical was 2.5 m/s. Whole measurement was carried out multiple times until temperature inside box stabilized. Setup can be seen in Figure 18.

Comparison of BER values with different bandwidths can be seen in Figure 19. It is noticeable that up to 32-QAM, BER values tended to stay below 10^{-5}. On the contrary, 64-QAM and higher modulations suffered from much higher BER. Figures 20 and 21 describe E_b/N_0 and EVM values for every M-QAM and bandwidth combination. By comparing these values to the reference, we can notice a slight decrease of E_b/N_0 and increase of EVM values. The most significant changes began at 128-QAM and progressed further. Communication ceased to work, so 2048-QAM was not measurable.

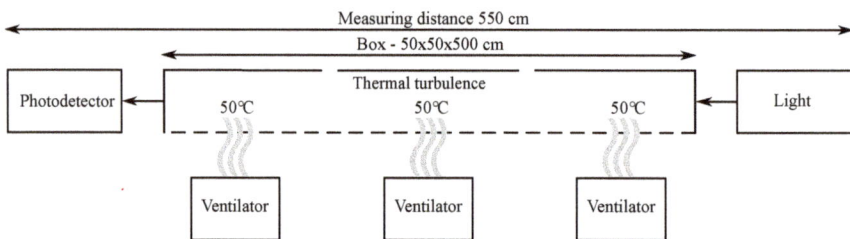

Figure 18. Octavia tail-light setup adjusted for scenario 2-thermal turbulence.

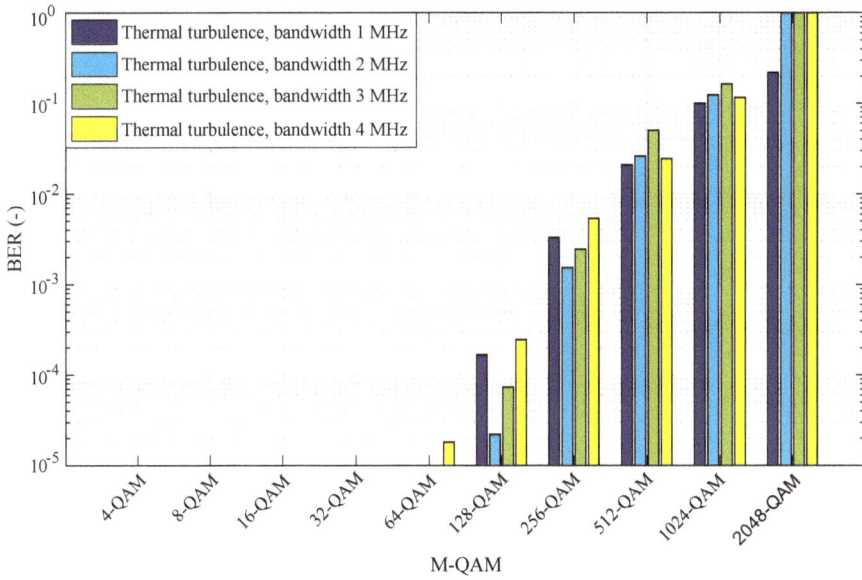

Figure 19. BER/distance relationship for Octavia tail-light with different M-QAM and bandwidths—scenario 2—thermal turbulence.

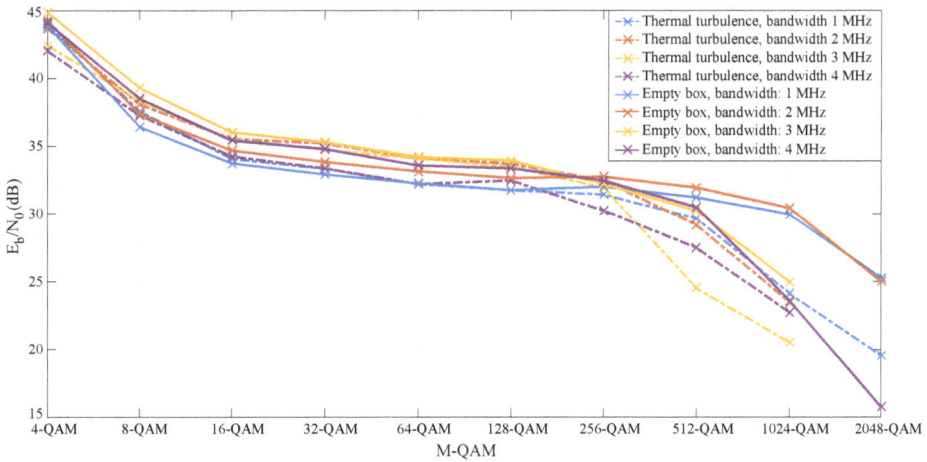

Figure 20. E_b/N_0/distance relationship for Octavia tail-light with different M-QAM and bandwidths—scenario 2—thermal turbulence.

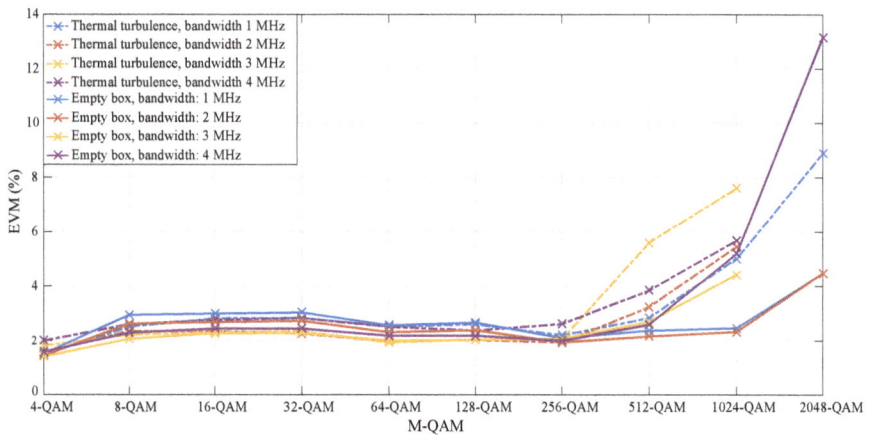

Figure 21. EVM/distance relationship for Octavia tail-light with different M-QAM and bandwidths—scenario 2—thermal turbulence.

4.3. Scenario 3-Rain 42 L/min

The first part of measurements with rain were carried out in modified box as well. In this scenario, we have removed the top part of the box and exchanged it with three water nozzles. The box was equipped with compressor which pumped water from the bottom of the box back into nozzles. The transmitter and receiver were located outside the box, which was opened from both sides. In this scenario, the water flow was set to 42 L/min. Setup can be seen on Figure 22. Figure 23 shows a photo of this setup.

Measurements were carried out in a room at 22–25 °C. Water temperature was stabilized at 22–25 °C before each measurement. Both transmitter and receiver were located 25 cm from open side of the box. This way, it was impossible for them to get fogged up. To investigate the influence of walls or partitions a series of tests without a box and in a completely dark room were carried out before. The box had minimal influence, as its construction was adjusted to prevent it. Transmitting light was precisely focused into the box to avoid interference, which could be caused by possible reflections. Incoming light was also focused into photodetector by planoconvex lens, as mentioned earlier. Box construction was spacious enough to avoid unnecessary reflections on running water. However, the influence of water particles on detector/transmitter itself will definitely be a topic of further research.

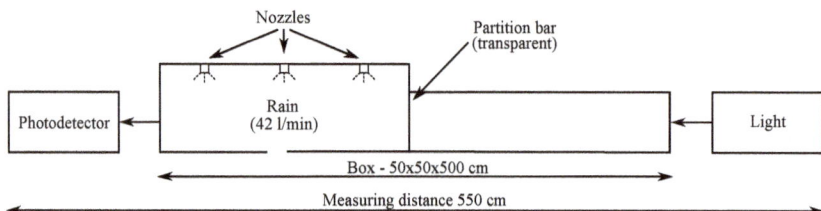

Figure 22. Octavia tail-light setup adjusted for scenario 3—Rain 42 L/min.

Figure 23. Photo of setup for scenario 3 and 4—rain.

Comparison of BER values with different bandwidths can be seen on Figure 24. This time, up to 8-QAM, BER stayed below $10e^{-5}$. On the contrary, 16-QAM and higher modulations suffer from much higher BER.

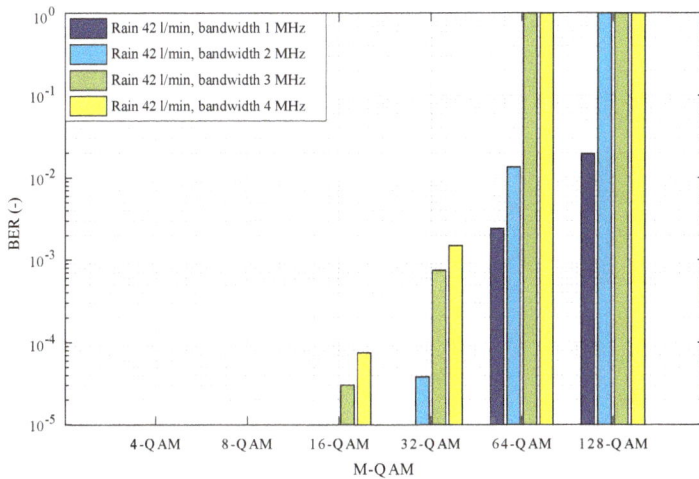

Figure 24. BER/distance relationship for Octavia tail-light with different M-QAM and bandwidths—scenario 3—Rain 42 L/min.

E_b/N_0 and EVM values are displayed on Figures 25 and 26. By comparing these values to reference setup, there was a significant drop in signal quality. For example, in case of 4-QAM modulation and 1 MHz bandwidth, SNR was nearly 10 dB lower than the reference. Significant decrease in signal quality was observed, which also led to lower maximal reachable modulation/bandwidth combination, which was 128-QAM/1 MHz.

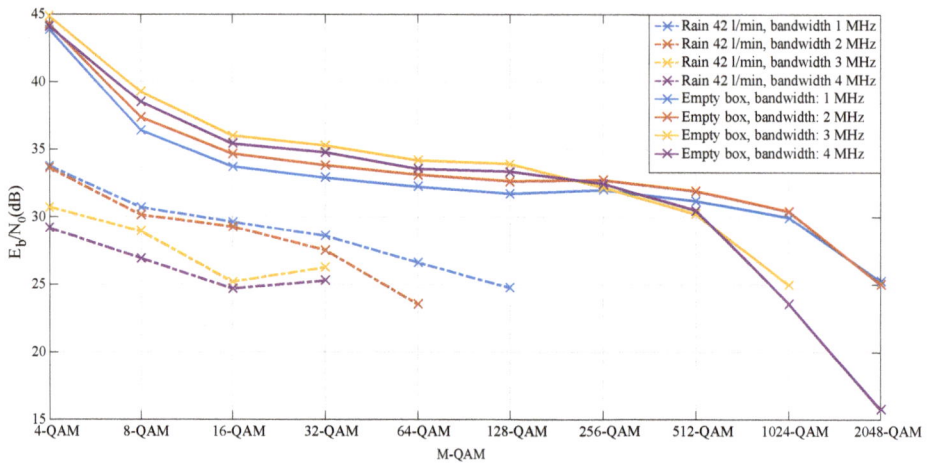

Figure 25. E_b/N_0 / distance relationship for Octavia tail-light with different M-QAM and bandwidths —scenario 3—Rain 42 L/min.

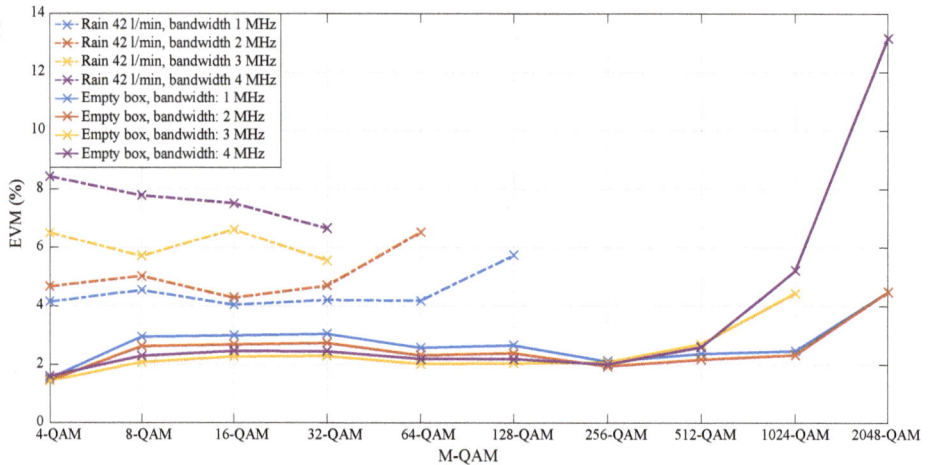

Figure 26. EVM/distance relationship for Octavia tail-light with different M-QAM and bandwidths —scenario 3—Rain 42 L/min.

This setup was also influenced by the drops of water on the sides of the box. Mainly the "separation wall" between box sections caused concerns. To analyze this concerning issue, we have carried out several measurements with different box setups, which can be seen on Figure 27. The first curve corresponds to the empty box, second one to the box with rain but without partition and the last one was the original measured setup. By comparing these values, there is a noticeable 3 dB increase in attenuation between second and third curve.

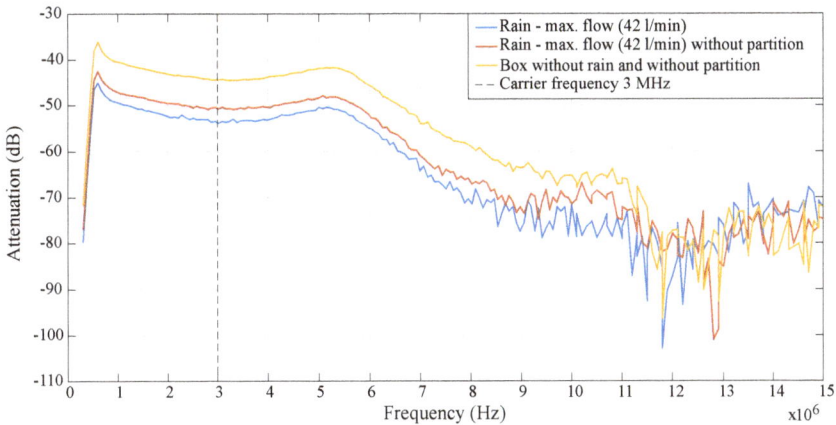

Figure 27. Attenuation characteristics of Octavia tail-light for scenario 3—Rain 42 L/min—comparison of intended setup and adjusted setup without partition.

4.4. Scenario 4 -Rain 22 L/min

The second part of measurements with rain were carried out in the modified setup from scenario 3. The main difference was a different water flow of 22 L/min. Setup can be seen on Figure 28.

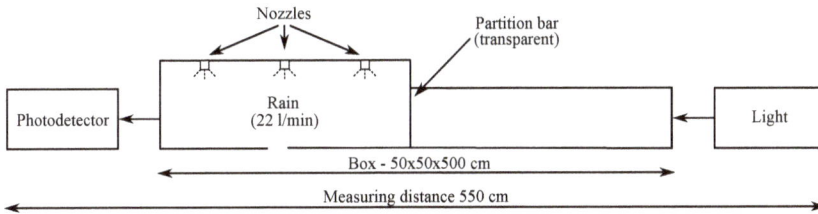

Figure 28. Octavia tail-light setup adjusted for scenario 4-Rain 22 L/min.

Comparison of BER values with different bandwidths can be seen on Figure 29. This time, up to 16-QAM, BER stayed below $10e^{-5}$. On the contrary, 32-QAM and higher modulations suffer from much higher BER.

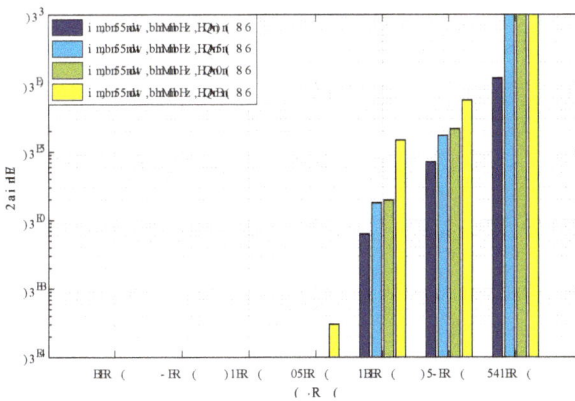

Figure 29. BER/distance relationship for Octavia tail-light with different M-QAM and bandwidths —scenario 4-Rain 22 L/min.

E_b/N_0 and EVM values (Figures 30 and 31) tended to follow the trend from previous scenario. This time, in case of 4-QAM modulation and 1 MHz bandwidth, SNR is nearly 7 dB lower than the reference scenario 1. Significant decrease in signal quality was observed, which also led to lower maximal reachable modulation/bandwidth combination, which was 256-QAM/1 MHz.

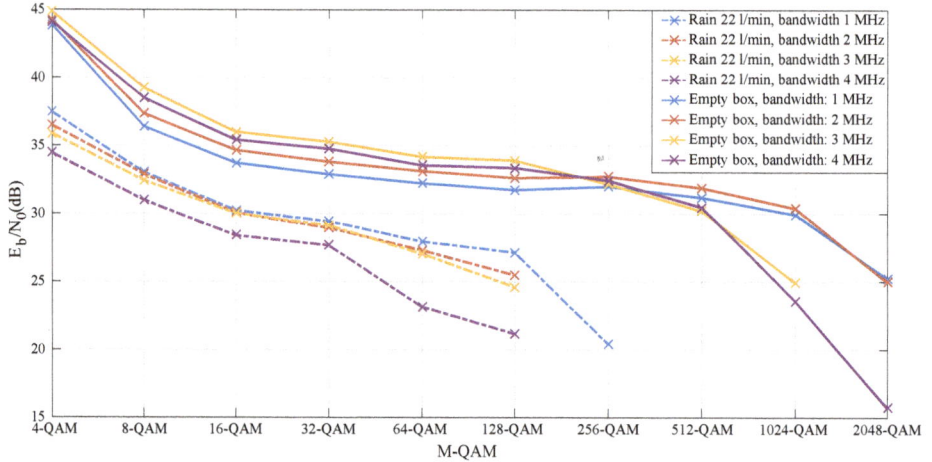

Figure 30. E_b/N_0 / distance relationship for Octavia tail-light with different M-QAM and bandwidths —scenario 4—Rain 22 L/min.

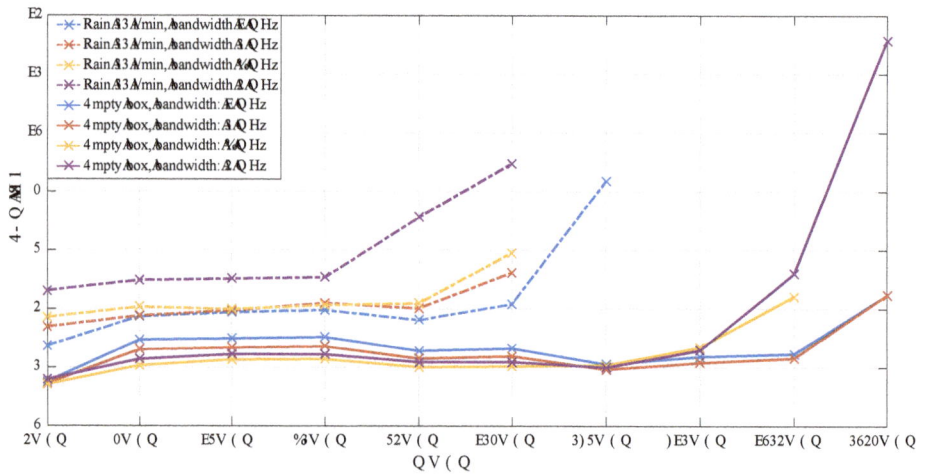

Figure 31. EVM/distance relationship for Octavia tail-light with different M-QAM and bandwidths —scenario 4—Rain 22 L/min.

Scenario 4 was also influenced by water drops. By carrying out the same test as in scenario 3, we have measured a 4 dB difference in attenuation, which was caused by "partition bar'". Measured values are displayed on Figure 32.

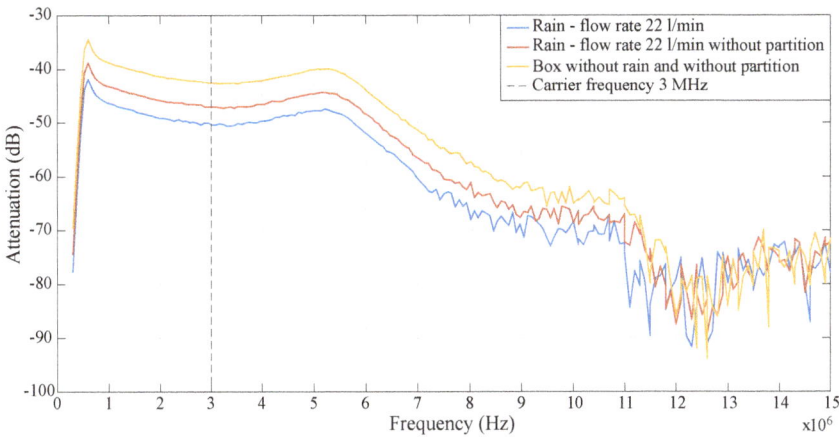

Figure 32. Attenuation characteristics of Octavia tail-light for scenario 4-Rain 22 L/min—comparison of intended setup and adjusted setup without partition.

4.5. Scenario 5-Fog

The last scenario consisted of a sealed box connected to the fog machine. This measurement differed from the others in that we did not always have the same conditions in the box, as fog inside gradually evaporated. That is why we had to add a 4 mW laser and optical power meter to our experiment (Figure 33). Figure 34 shows a photo of this setup. During measurements, we have monitored and gathered optical power levels to use them as reference values. Due to the slow fog evaporation, higher modulations gradually reached the desired BER threshold of 10^{-3}. As soon as it happened, we carried out our main measurements. In this case we have measured all bandwidths at once, to preserve credibility. Figure 35 describes a slow dissipation of fog inside box with gathered laser optical power levels and BER values for 1 MHz bandwidth.

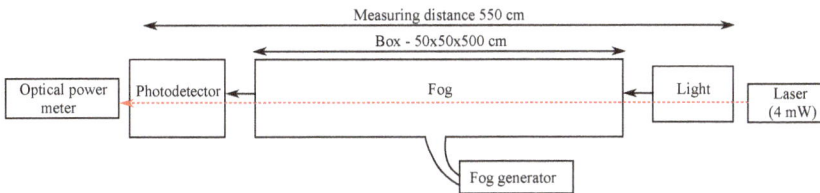

Figure 33. Octavia tail-light setup adjusted for scenario 5–fog.

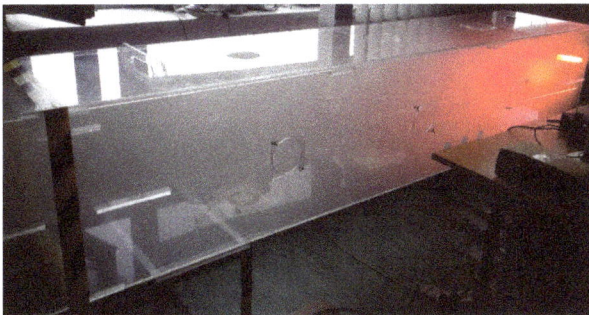

Figure 34. Photo of setup for scenario 5-fog.

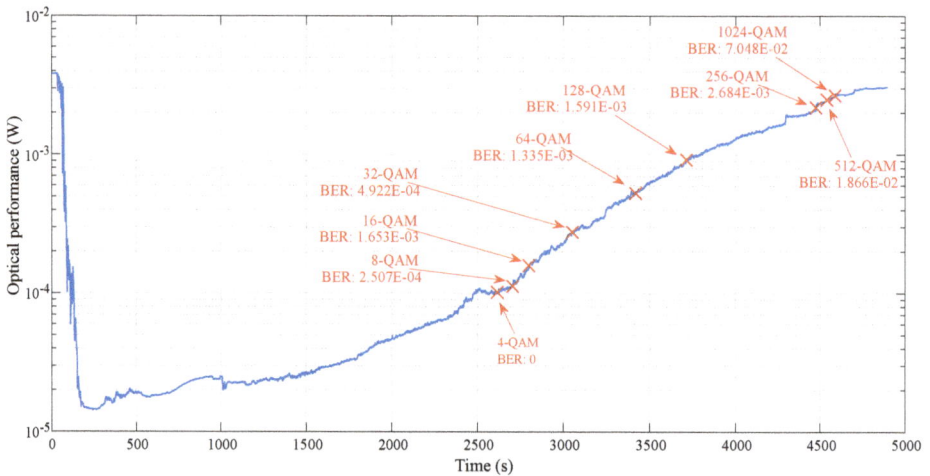

Figure 35. Fog dissipation with laser power levels and modulations with appropriate BER values.

The Laser itself was mainly used to exactly determine how fast the fog dissipated (concentration) and if it dissipated similarly in multiple measurements, which proved to be true. High directionality of laser beam and planoconvex lens on photodetector helped us to avoid any unnecessary effects of second light source. As mentioned the box was big enough to host both devices, which can work independently. In this case, the laser was the available device which we already had and could be used.

Table 4 consists of measured optical power level thresholds, after passing through fog. It is visible that at least a quarter of transmitted optical power is needed for higher modulation formats (128-QAM+) to work. That is why following the figures display minimal values for successful initiation of communication between transmitter and receiver for multiple modulation formats.

Table 4. Minimal optical power levels of laser for multiple bandwidth and modulations measured after passing through fog.

M-QAM	Bandwidth			
-	**1 MHz**	**2 MHz**	**3 MHz**	**4 MHz**
4-QAM	101 μW	95 μW	97 μW	93 μW
8-QAM	112 μW	135 μW	175 μW	189 μW
16-QAM	154 μW	205 μW	250 μW	276 μW
32-QAM	275 μW	308 μW	463 μW	550 μW
64-QAM	525 μW	780 μW	930 μW	951 μW
128-QAM	905 μW	1.095 mW	1.399 mW	1.550 mW
256-QAM	2.180 mW	2.394 mW	2.545 mW	2.585 mW
512-QAM	2.475 mW	2.522 mW	2.735 mW	2.930 mW
1024-QAM	2.640 mW	2.955 mW	3.120 mW	3.172 mW

Measured BER values are displayed on Figure 36. This figure consists of all modulation formats, where BER values were better than 1. However, it is necessary to compare these values to Table 4, as different optical power levels were needed for every modulation scheme. For example, at least a quarter of transmitted optical power had to be received for 128-QAM/2MHz combination to work. E_b/N_0 and EVM values displayed on Figures 37 and 38 also must be compared to Table 4.

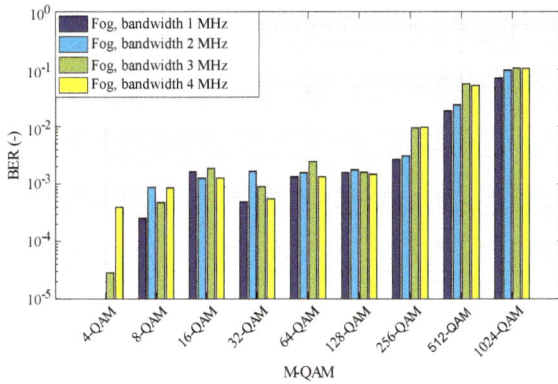

Figure 36. BER/distance relationship for Octavia tail-light with different M-QAM and bandwidths—in scenario 5—fog.

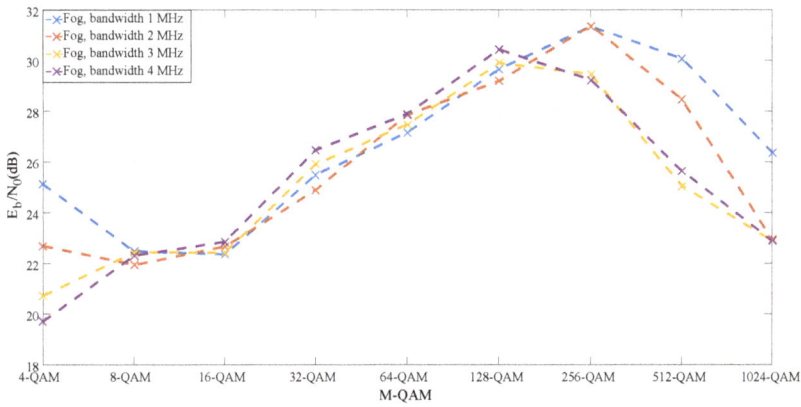

Figure 37. E_b/N_0 / distance relationship for Octavia tail-light with different M-QAM and bandwidths —in scenario 5—fog.

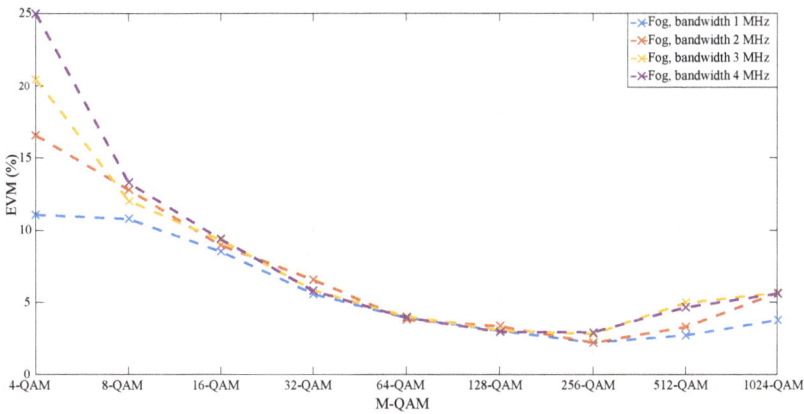

Figure 38. EVM/distance relationship for Octavia tail-light with different M-QAM and bandwidths —scenario 5—fog.

From all the meteorological phenomena, fog had the greatest influence on visible light communication and will be the hardest to overcome. Figure 39 displays attenuation characteristics for different laser power levels. For example, at 1 mW output power, attenuation increased by 63 dB in comparison with reference values.

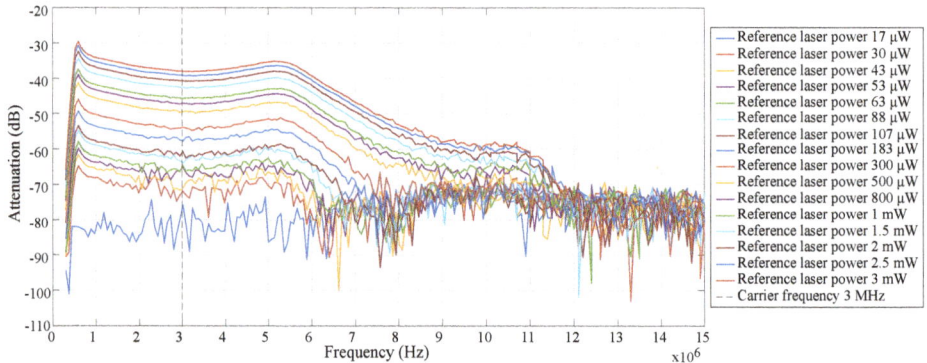

Figure 39. Attenuation characteristics of Octavia tail-light during fog dissipation –in scenario 5 - fog.

5. Future Research and Discussion

During testing we have encountered several shortcomings, we will try to overcome in the future. As our prototype is designed with the highest possible modularity in mind, we want to aim for better parts especially tailored for our needs. We have identified amplifier nonlinearity and photodetector as our biggest limitation. We were able to acquire two APD modules from Hamamatsu Photonics Japan, which will be tested, evaluated and possibly incorporated into our prototype [103,104]. There are also multiple companies we are currently in contact with to manufacture custom amplifiers such as Mini Circuits and Ophir RF.

Figures 6 and 13 show trend of increasing attenuation below 1 MHz and above 5 MHz. It is caused either by insufficient optimization of inbuild pre or post equalizers or by an amplifier in combination with insufficient impedance matching. Nonlinearity at 1 MHz to 5 MHz did not cause any significant changes to transmission quality. However we are currently working on optimization and have ordered a number of specialized parts.

We would like to propose a new concept of software defined optics (SDO). SDO is a visible light communication system, in which the critical part of signal processing is implemented by software programmable circuits. As a result, a variety of software changes can be used to swiftly modify the communication system, such as used modulation, channel coding etc. The transmitter consists of D/A converter and appropriate transmitter (LED or laser diode), which is modulated by a bias-tee. The receiver consists of a photodiode or camera and A/D converter.

The main advantage of our experiment was deployment of commercially available lights as transmitters. We avoided excessive modifications of original designs and tested them to their thresholds. However, we estimate, that modification of an Octavia tail-light LED cluster might increase effective maximal communication distance by 20%, as original LED matrixes tend to be of lower quality.

Implementation of adaptive modulation and/or channel width would allow our setup to dynamically react to different conditions. We have also estimated that channel equalization would improve whole prototype significantly. Channel equalization implementation will be a topic of our immediate research, which will follow this paper shortly.

Next revision of our software will replace simpler QAM with full-fledged OFDM, which will also significantly improve our prototype. The next generation of our LabVIEW SDR VLC implementation is currently under development and will be a topic of further papers.

Another advantage of our prototype is its software based on virtual instrumentation. LabVIEW SDR offers us a reliable and highly modular platform, which we can easily modify, adjust or move. Whole software was frequently moved between stationary desktop computer and powerful laptop, so we were able to test it in different labs with custom equipment, such as special box mentioned earlier. Main advantage lays in LabVIEW modularity. We can implement multiple functions very fast, such as addition of channel coding or equalization, which is a huge advantage in debug phase. The final implementation of our prototype will be minimalized and optimized. We estimate, that third generation of our software (second generation of OFDM) will run natively on FPGA.

Each measurements/tests were carried out multiple times and mainly independently. Each scenario was specially designed to represent the most pressing problems in VLC. Variable natural conditions and their compensation are a topic for multiple teams and their research [6,9,37,105]. As fog is the most concerning problem, several teams tried to compensate its impact [47,106,107]. A sandstorm, which was not tested in our paper, is considered as another concerning topic, however according to simulations, its characteristics is very similar to fog and rain [108].

We have also acquired a number of blue-light filters from Thorlabs. According to some papers, a system which uses these filters will have wider useable bandwidth but might suffer from shorter communication distance. Analysis of these parameters will be a topic of further research. As mentioned earlier, our Octavia taillight setup used plano-convex lens, which were already available at our university. We have also acquired another set of plano-convex and Fresnel lens which will be also a topic of further research.

Setup with indoor ceiling light could be enhanced with VLC positioning system, as there are a number of proposed implementations [109]. However primary purpose is data transmissions, so positioning would be only a supplementary function, which should not interfere with primary purpose or limit measured parameters. Some teams even reached up to 95 % accuracy in their experiments. OFDM, or more specifically OFDMA can be used for data transmissions as well as positioning. Positioning methods are usually based on trilateration model, which needs at least three transmitters with known location. However, we have used only one ceiling light for prototyping, so positioning would not be possible. This area will be a topic of further research [110–114].

In case of VLC, most teams are mainly focused on achieving longer communication distances or higher transmit speeds. Security was pushed sideways in favor of other areas. However we agree with colleagues, that non-line-of-sight (NLoS) or even line-of-sight (LoS) eavesdropping is possible without sturdy security, which is not covered enough in IEEE standardization. Possible NLoS eavesdropping will be a topic for further research, as we are just entering this area [115].

Many teams tend to use their own LED matrixes or custom designed lights. However, that is concerning, as quality of deployed LEDs in commercially available products tend to vary significantly. Manufacturers often use lower quality LEDs, because even a small difference in price will make a huge impact in higher quantities. That's why we split our research into two branches. We begun our research on commercially available lights, as they are "target hardware". However we are also running a number of test on multiple LEDs of different qualities, and are looking for best price/performance ratio. This area will be also a topic of further research [7,79–82].

Technical University of Ostrava has its own newly developed testing polygon called "BroadbandLIGHT". BroadbandLIGHT is situated next to faculty of electrical engineering and computer science. It consists of 20 lamp posts and central management located in laboratory EB418 (see Figure 40). The whole system is prepared for future implementation of VLC technology. It is a unique polygon that is oriented at potential customers, demonstrations of functionality or long-term measurements. It is a next step for incorporation of smart technologies (smart city, IoT, Industry 4.0) into lighting and can be used for either indoor or outdoor setups. Detailed description and possible implementation will be discussed in further papers.

A new kind of hybrid system is currently in development by our team. It is based on powerline, which will be connected directly to ceiling light. The return channel is based on power efficient

Bluetooth technology. It Is estimated that whole system should be much more energy efficient. It is based on the premise that most of the users download much more than they upload, which will result in lower battery drain.

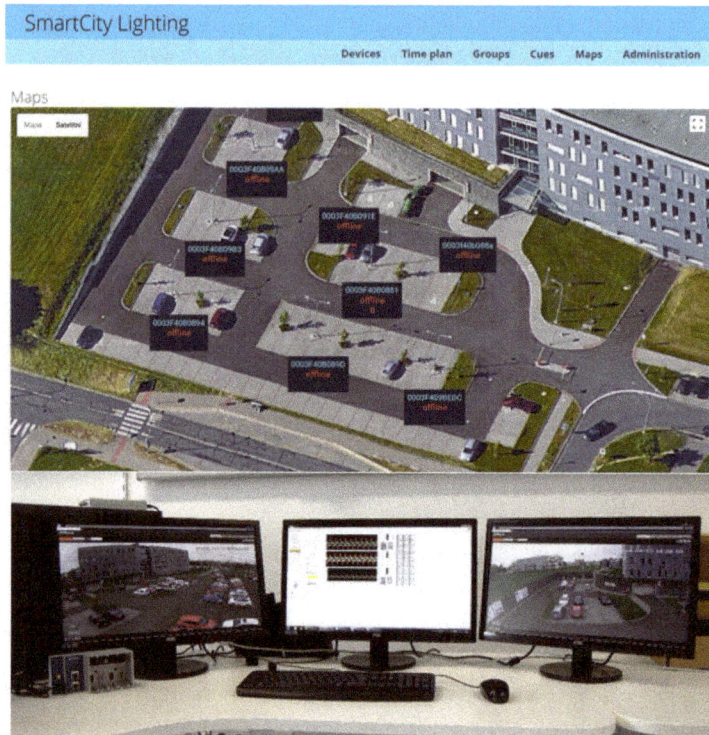

Figure 40. BroadbandLIGHT testing polygon implementation and management center.

Table 5 summarizes multiple tests with different configurations, including ours. We have included information about modulation, detector, maximal reachable transmit speed and even maximal reachable distance.

Next version of our prototype will feature advanced signal processing methods, which we already investigated multiple times in different areas [116–121].

Table 5. Comparison of different VLC experiments.

Year	Main Author	Source	Distance	Detector	Modulation Type	Transmit Speed
2016	Wen-Hsuan Shen [22]	Tail-light 1157 (12 V, red LED)	45 m	Photodiode Thorlabs PDA100A (10 mm × 10 mm)	PWM (OFDM)	45–55 Mbps
2016	Yuki Goto [30]	Red LED matrix (4 × 5; 0.72 V, 14.5 MHz)	1.5 m	OCI camera (Focal distance of the lens 50 mm)	4–256 QAM (OFDM)	45 Mbps 50 Mbps 55 Mbps
2014	Isamu Takai [22]	Red LED matrix (10 × 10; 4 W, 55 MHz)	7.78 m	High-speed camera (Focal distance of the lens 12.5 mm)	Pulse Width modulation (PWM)	10 Mpbs
2014	Takaya Yamazato [122]	Red LED matrix (32 × 32)	45 m	High-speed camera (CMOS sensor)	Pulse Width modulation (PWM)	10 Mpbs
2016	Yoshihito Imai [123]	Red LED (OS5RKA5B61P)	1–8 m	High-speed camera XCG-V60E, SONY (Focal distance of the lens 8–48 mm)	Pulse Width modulation (PWM)	84 bps
2017	Takaya Yamazato [124]	Red LED matrix (32 × 32)	30–70 m	High-speed image sensor and OCI sensor	Pulse Width modulation (PWM)	55 Mbps
2019	Radek Martinek	Phillips Fortimo DLM 300 44 W/840 Gen3	3.25 m from center	Thorlabs PDA-36A	QAM modulation	2 Mbps
2019	Radek Martinek	Skoda Octavia Tail-light	5.5 m	Thorlabs PDA-36A	QAM modulation	28 Mbps

6. Conclusions

In this paper, we have presented results of our ongoing research. Our VLC prototype based on NI LabVIEW SDR and commercially available lights is introduced. First part of our experiment was carried out on Phillips indoor ceiling light. We have analyzed maximal possible transmit speeds at different distances from center, located directly below transmitter. The maximal reachable distance was 325 m, when using 4-QAM and 1 MHz bandwidth. Transmit speed was at least 2 Mbps. With wider channels and higher modulations (8-QAM to 256-QAM), maximal reachable distance decreased rapidly.

The second part was based on Octavia tail-light, which was tested under different natural conditions. A maximal measuring distance was constant 550 cm. A special box was inserted between transmitter (tail-light) and receiver (photodetector), which simulated different conditions such as: thermal turbulence, rain and fog. A number of figures with real measured data was presented and commented accordingly. We have also identified weak spots, which will be adjusted and possibly researched in future publications. From measured data, fog has the greatest impact on communication, as it caused. On the other side, thermal turbulence had the lowest impact. The highest reachable transmit speed was 28 Mbps for combination of 256-QAM and 4MHz, which was reachable only in the empty box.

Weak spots were identified and thoroughly discussed. A number of possible upgrades were mentioned, focusing on custom parts, which will significantly improve whole setup.

Author Contributions: Conceptualization, R.M. and L.D.; Methodology, R.M.; Software, L.D. and R.M.; Validation, R.M., L.D. and R.J.; Formal Analysis, R.M. and L.D.; Investigation, R.M. and R.J.; Resources, R.M.; Data Curation, L.D.; Writing—Original Draft Preparation, L.D.; Writing—Review & Editing, R.M. and R.J.; Visualization, R.J.; Supervision, R.M.; Project Administration, R.M.; Funding Acquisition, R.M.

Funding: This article was supported by the Ministry of Education of the Czech Republic (Project No. SP2019/85). This work was supported by the European Regional Development Fund in the Research Centre of Advanced Mechatronic Systems project, project number CZ.02.1.01/0.0/0.0/16_019/0000867 within the Operational Programme Research, Development and Education. This work was supported by the European Regional Development Fund in A Research Platform focused on Industry 4.0 and Robotics in Ostrava project, CZ.02.1.01/0.0/0.0/17_049/0008425 within the Operational Programme Research, Development and Education.

Conflicts of Interest: The authors declare no conflicts of interest.

References

1. Rajagopal, S.; Roberts, R. D.; Lim, S. K. IEEE 802.15. 7 visible light communication: Modulation schemes and dimming support. *IEEE Commun. Mag.* **2012**, *50*, 72–82. [PubMed] [CrossRef]
2. Le Minh, H.; O'Brien, D.; Faulkner, G.; Zeng, L.; Lee, K.; Jung, D.; Oh, T.; Won, E.T. 100-Mb/s NRZ visible light communications using a postequalized white LED. *IEEE Photonics Technol. Lett.* **2009**, *21*, 1063–1065. [PubMed] [CrossRef]
3. Biton, C.; Arnon, S. Improved multiple access resource allocation in visible light communication systems. *Opt. Commun.* **2018**, *424*, 98–102. [PubMed] [CrossRef]
4. Cailean, A.M.; Dimian, M. Current challenges for visible light communications usage in vehicle applications: A survey. *IEEE Commun. Surveys Tutor.* **2017**, *19*, 2681–2703. [PubMed] [CrossRef]
5. Bekhrad, P.; Leitgeb, E.; Ivanov, H. Benefits of visible light communication in car-to-car communication. In Proceedings of the of the Fiber Lasers and Glass Photonics: Materials through Applications, Swansea, UK, 22–26 April 2018; Volume 10683, p. 106833A. [CrossRef] [CrossRef]
6. Singh, G.; Srivastava, A.; Bohara, V.A. On Feasibility of VLC Based Car-to-Car Communication Under Solar Irradiance and Fog Conditions. In Proceedings of the 1st International Workshop on Communication and Computing in Connected Vehicles and Platooning, New Delhi, India, 29 October 2018; pp. 1–7. [PubMed] [CrossRef]

7. Matus, V.; Maturana, N.; Azurdia-Meza, C.A.; Montejo-Sanchez, S.; Rojas, J. Hardware design of a prototyping platform for vehicular VLC using SDR and exploiting vehicles CAN bus. In Proceedings of the of the 2017 First South American Colloquium on Visible Light Communications (SACVLC), Santiago, Chile, 13 November 2017; pp. 1–4. [PubMed]

8. Luo, P.; Ghassemlooy, Z.; Le Minh, H.; Bentley, E.; Burton, A.; Tang, X. Fundamental analysis of a car to car visible light communication system. In Proceedings of the of the 2014 9th International Symposium on Communication Systems, Networks & Digital Sign (CSNDSP), Manchester, UK, 23–25 July 2014; pp. 1011–1016. [PubMed] [CrossRef]

9. Kim, J.Y.; Park, B.S.; Choi, H.S.; Kim, S.E.; Moon, I.; Lee, C.G. Effect of interferences on indoor visible light car-to-car communication systems. In Proceedings of the Optical Modelling and Design IV, Brussels, Belgium, 3–7 April 2016; Volume 9889, p. 98891U. [CrossRef] [CrossRef]

10. Arnon, S. Optimised optical wireless car-to-traffic-light communication. *Trans. Emerg. Telecommun. Technol.* **2014**, *25*, 660–665. [PubMed] [CrossRef]

11. Jeong, J.; Lee, C.G.; Moon, I.; Kang, M.; Shin, S.; Kim, S. Receiver angle control in an infrastructure-to-car visible light communication link. In Proceedings of the 2016 IEEE Region 10 Conference (TENCON), Marina Bay Sands, Singapore, 22–25 November 2016; pp. 1957–1960. [PubMed] [CrossRef]

12. Deng, P.; Kavehrad, M. Real-time software-defined single-carrier QAM MIMO visible light communication system. In Proceedings of the 2016 Integrated Communications Navigation and Surveillance (ICNS), Herndon, VA, USA, 19–21 April 2016; pp. 5A3-1–5A3-11. [PubMed] [CrossRef]

13. Tsonev, D.; Videv, S.; Haas, H. Light fidelity (Li-Fi): Towards all-optical networking. In Proceedings of the Broadband Access Communication Technologies VIII, San Francisco, CA, USA, 1–6 February 2014; p. 900702. [CrossRef] [CrossRef]

14. Saini, H. Li-Fi (Light Fidelity)-The future technology In Wireless communication. *J. Comput. Appl.* **2016**, *7*, 13–15. [PubMed]

15. Bao, X.; Yu, G.; Dai, J.; Zhu, X. Li-Fi: Light fidelity-a survey. *Wireless Netw.* **2015**, *21*, 1879–1889. [CrossRef] [CrossRef]

16. Haas, H.; Yin, L.; Wang, Y.; Chen, C. What is lifi? *J. Lightwave Technol.* **2016**, *34*, 1533–1544. [CrossRef] [CrossRef]

17. Sharma, R.R.; Sanganal, A. Li-Fi Technology: Transmission of data through light. *Int. J. Comput. Technol. Appl.* **2014**, *5*, 150. [PubMed]

18. Nivrutti, D. V.; Nimbalkar, R. R. Light-Fidelity: A Reconnaissance of Future Technology. *Int. J. Adv. Res. Comput. Sci. Softw. Eng.* **2013**, *3*. [PubMed]

19. Sarkar, A.; Agarwal, S.; Nath, A. Li-fi technology: Data transmission through visible light. *Int. J. Adv. Res. Comput. Sci. Manag. Stud.* **2015**, *3*. [PubMed]

20. Haas, H. LiFi: Conceptions, misconceptions and opportunities. In Proceedings of the 2016 IEEE Photonics Conference (IPC), Waikoloa, HI, USA, 2–6 October 2016; pp. 680–681. [PubMed] [CrossRef]

21. Jaiswal, N.S.; Chopade, P.S. Review of Li-Fi technology: New future technology-light bulb to access the internet! *Int. J. Sci. Eng. Res.* **2013**, *4*, 36–40.

22. Shen, W.H.; Tsai, H.M. Testing vehicle-to-vehicle visible light communications in real-world driving scenarios. In Proceedings of the 2017 IEEE Vehicular Networking Conference (VNC), Torino, Italy, 27–29 November 2017; pp. 187–194. [PubMed] [CrossRef]

23. Stewart, R.W.; Barlee, K.W.; Atkinson, D.S.; Crockett, L.H. *Software Defined Radio Using MATLAB & Simulink and the RTL-SDR*; Strathclyde Academic Media: Glasgow, UK, 2015.

24. Al Wohaishi, M.; Zidek, J.; Martinek, R. Analysis of M state digitally modulated signals in communication systems based on SDR concept. In Proceedings of the 6th IEEE International Conference on Intelligent Data Acquisition and Advanced Computing Systems, Prague, Czech Republic, 15–17 September 2011; pp. 171–175. [PubMed] [CrossRef]

25. Rahaim, M.; Miravakili, A.; Borogovac, T.; Little, T. D. C.; Joyner, V. Demonstration of a software defined visible light communication system. In the 17th Annual International Conference on Mobile Computing and Networking, Mobicom2011, Las Vegas, NV, USA, 19–23 September 2011. [PubMed]

26. Hussain, W.; Ugurdag, H.F.; Uysal, M. Software defined VLC system: Implementation and performance evaluation. In Proceedings of the 2015 4th International Workshop on Optical Wireless Communications (IWOW), Istanbul, Turkey, 7–8 September 2015; pp. 117–121. [PubMed] [CrossRef]

27. Philips. Philips Fortimo LED Downlight Module System. Phillips Fortimo DLM 300 44 W/840 Gen3 Datasheet. Available online: http://media.futureelectronics.com/semiconductors/lighting-solutions/led-light-modules/fortimo_dlm_twist.pdf?m=Q3SoAq.pdf (accessed on 28 February 2019). [PubMed]

28. Thorlabs. PDA36A Operating Manual-Switchable Gain, Amplified Silicon Detector. PDA36A-EC Datasheet. Avaialble online: http://physics-astronomy-manuals.wwu.edu/Thorlabs%20PDA36A%20Detector%20Manual.pdf (access on 28 February 2019) [PubMed]

29. Masini, B.; Bazzi, A.; Zanella, A. A survey on the roadmap to mandate on board connectivity and enable V2V-based vehicular sensor networks. *Sensors* **2018**, *18*, 2207. [CrossRef] [CrossRef] [PubMed]

30. Goto, Y.; Takai, I.; Yamazato, T.; Okada, H.; Fujii, T.; Kawahito, S.; Shintaro, A.; Tomohiro Yendo, T.; Kamakura, K. A new automotive VLC system using optical communication image sensor. *IEEE Photonics J.* **2016**, *8*, 1–17. [PubMed] [CrossRef]

31. Raza, N.; Jabbar, S.; Han, J.; Han, K. Social vehicle-to-everything (V2X) communication model for intelligent transportation systems based on 5G scenario. In Proceedings of the 2nd International Conference on Future Networks and Distributed Systems, Amman, Jordan, 26–27 June 2018; p. 54. [PubMed] [CrossRef]

32. Kinoshita, M.; Yamazato, T.; Okada, H.; Fujii, T.; Arai, S.; Yendo, T.; Kamakura, K. Motion modeling of mobile transmitter for image sensor based I2V-VLC, V2I-VLC, and V2V-VLC. In Proceedings of the 2014 IEEE Globecom Workshops (GC Wkshps), Austin, TX, USA, 8–12 December 2014; pp. 450–455. [PubMed] [CrossRef]

33. Kim, B. W.; Jung, S.Y. Vehicle positioning scheme using V2V and V2I visible light communications. In Proceedings of the 2016 IEEE 83rd Vehicular Technology Conference (VTC Spring), Nanjing, China, 20 September 2015; pp. 1–5. [PubMed] [CrossRef]

34. Vaezi, M.; Ding, Z.; Poor, H.V. *Multiple Access Techniques for 5G Wireless Networks and Beyond*; Springer: Berlin, Germany, 2019. [PubMed]

35. Boban, M.; Kousaridas, A.; Manolakis, K.; Eichinger, J.; Xu, W. Connected roads of the future: Use cases, requirements, and design considerations for vehicle-to-everything communications. *IEEE Veh. Technol. Mag.* **2016**, *13*, 110–123. [PubMed] [CrossRef]

36. Cailean, A.M.; Cagneau, B.; Chassagne, L.; Popa, V.; Dimian, M. A survey on the usage of DSRC and VLC in communication-based vehicle safety applications. In Proceedings of the 2014 IEEE 21st Symposium on Communications and Vehicular Technology in the Benelux (SCVT), Delft, The Netherlands, 10 November 2014; pp. 69–74. [PubMed] [CrossRef]

37. Kim, Y.H.; Cahyadi, W.A.; Chung, Y.H. Experimental demonstration of LED-based vehicle to vehicle communication under atmospheric turbulence. In Proceedings of the 2015 International Conference on Information and Communication Technology Convergence (ICTC), Jeju Island, Korea, 18–30 October 2015; pp. 1143–1145. [PubMed] [CrossRef]

38. Campolo, C.; Molinaro, A.; Scopigno, R. From today's VANETs to tomorrow's planning and the bets for the day after. *Veh. Commun.* **2015**, *2*, 158–171. [CrossRef] [CrossRef]

39. Yamazato, T. Overview of visible light communications with emphasis on image sensor communications. In Proceedings of the 2017 23rd Asia-Pacific Conference on Communications (APCC), Perth, WA, Australia, 11–13 December 2017; pp. 1–6. [PubMed] [CrossRef]

40. Yamazato, T. Image Sensor Communications for future ITS. In Proceedings of the Signal Processing in Photonic Communications (pp. SpW2G-6). Optical Society of America, Zurich, Switzerland, 2–5 July 2018. [CrossRef] [CrossRef]

41. Tsado, Y.; Lund, D.; Gamage, K.A. Resilient communication for smart grid ubiquitous sensor network: State of the art and prospects for next generation. *Comput. Commun.* **2015**, *71*, 34–49. [CrossRef] [CrossRef]

42. Tareen, W.U.K.; Mekhilef, S.; Nakaoka, M. A transformerless reduced switch counts three-phase APF-assisted smart EV charger. In Proceedings of the 2017 IEEE Applied Power Electronics Conference and Exposition (APEC), Tampa, FL, USA, 26–30 March 2017; pp. 3307–3312. [PubMed] [CrossRef]

43. Roche, R.; Berthold, F.; Gao, F.; Wang, F.; Ravey, A.; Williamson, S. A model and strategy to improve smart home energy resilience during outages using vehicle-to-home. In Proceedings of the 2014 IEEE International Electric Vehicle Conference (IEVC), Florence, Italy, 17–19 December 2014; pp. 1–6. [PubMed] [CrossRef]

44. Turker, H.; Bacha, S. Optimal Minimization of Plug-in Electric Vehicle Charging Cost with Vehicle-to-Home and Vehicle-to-Grid concepts. *IEEE Trans. Veh. Technol.* **2018** *67*, 10281–10292. [PubMed] [CrossRef]

45. Yamazato, T. V2X communications with an image sensor. *J. Commun. Inf. Netw.* **2017**, *2*, 65–74. [CrossRef] [CrossRef]

46. Mare, R.M.; Marte, C.L.; Cugnasca, C.E. Visible light communication applied to intelligent transport systems: An Overview. *IEEE Latin Am.Trans.* **2016**, *14*, 3199–3207. [PubMed] [CrossRef]

47. Kim, Y.H.; Cahyadi, W.A.; Chung, Y.H. Experimental demonstration of VLC-based vehicle-to-vehicle communications under fog conditions. *IEEE Photonics J.* **2015**, *7*, 1–9. [PubMed] [CrossRef]

48. Kinoshita, M.; Yamazato, T.; Okada, H.; Fujii, T.; Arai, S.; Yendo, T.; Kamakura, K. Channel fluctuation measurement for image sensor based I2v-VLC, V2i-VLC, and V2v-VLC. In Proceedings of the 2014 IEEE Asia Pacific Conference on Circuits and Systems (APCCAS), Okinawa, Japan, 17–20 November 2014; pp. 332–335. [PubMed] [CrossRef]

49. Arena, F.; Pau, G. An Overview of Vehicular Communications. *Future Internet* **2019**, *11*, 27. [CrossRef] [CrossRef]

50. Goldman-Shenhar, C.V.; Friedland, Y.S.; Riess, E.; Philosof, T.; Tsimhoni, O. Vehicle-to-Pedestrian-Communication Systems and Methods for Using the Same. U.S. Patent No. 9,881,503, U.S. Patent 30 January 2018.

51. Viriyasitavat, W.; Boban, M.; Tsai, H. M.; Vasilakos, A. Vehicular communications: Survey and challenges of channel and propagation models. *IEEE Veh. Technol. Mag.* **2015**, *10*, 55–66. [PubMed] [CrossRef]

52. Wang, Y.; Sheikh, O.; Hu, B.; Chu, C.C.; Gadh, R. Integration of V2H/V2G hybrid system for demand response in distribution network. In Proceedings of the 2014 IEEE International Conference on Smart Grid Communications (SmartGridComm), Venice, Italy, 3–6 November 2014; pp. 812–817. [PubMed] [CrossRef]

53. Zhao, L.; Aravinthan, V. Strategies of residential peak shaving with integration of demand response and V2H. In Proceedings of the 2013 IEEE PES Asia-Pacific Power and Energy Engineering Conference (APPEEC), Kowloon, Hong Kong, 11 December 2013; pp. 1–5. [PubMed] [CrossRef]

54. Guille, C.; Gross, G. A conceptual framework for the vehicle-to-grid (V2G) implementation. *Energy Policy* **2009**, *37*, 4379–4390. [CrossRef] [CrossRef]

55. Ota, Y.; Taniguchi, H.; Nakajima, T.; Liyanage, K.M.; Baba, J.; Yokoyama, A. Autonomous distributed V2G (vehicle-to-grid) satisfying scheduled charging. *IEEE Trans. Smart Grid* **2012**, *3*, 559–564. [PubMed] [CrossRef]

56. Alam, M. Vehicle-to-Everything (V2X) Technology Will Be a Literal Life Saver But What Is It?, 2016. Available online: http://eecatalog.com/automotive/2016/05/19/vehicle-to-everything-v2x-technology-will-be-a-literal-life-saver-but-what-is-it/ (access on: 28 February 2019) [PubMed]

57. Ng, X.W.; Chung, W.Y. VLC-based medical healthcare information system. *Biomed. Eng. Appl. Basis Commun.* **2012**, *24*, 155–163. [CrossRef] [CrossRef]

58. Ding, W.; Yang, F.; Yang, H.; Wang, J.; Wang, X.; Zhang, X.; Song, J. A hybrid power line and visible light communication system for indoor hospital applications. *Comput. Ind.* **2015**, *68*, 170–178. [CrossRef] [CrossRef]

59. An, J.; Chung, W.Y. A novel indoor healthcare with time hopping-based visible light communication. In Proceedings of the 2016 IEEE 3rd World Forum on Internet of Things (WF-IoT), Reston, VA, USA, 12–14 December 2016; pp. 19–23. [PubMed] [CrossRef]

60. Song, J.; Ding, W.; Yang, F.; Yang, H.; Wang, J.; Wang, X.; Zhang, X. Indoor hospital communication systems: An integrated solution based on power line and visible light communication. In Proceedings of the 2014 IEEE Faible Tension Faible Consommation, Monaco, Monaco, 4–6 May 2014; pp. 1–6. [PubMed] [CrossRef]

61. Tagliaferri, D.; Capsoni, C. SNIR predictions for on-aircraft VLC systems. In Proceedings of the 2016 International Conference on Broadband Communications for Next Generation Networks and Multimedia Applications (CoBCom), Graz, Austria, 14–16 September 2016 ; pp. 1–7. [PubMed] [CrossRef]

62. Png, L.C.; Lim, S.X.; Rajamohan, A.; Chan, B.W.; Hazman, F.A. Designs of VLC transceiver circuits for reading light transmission of high-quality audio signals on commercial airliners. In Proceedings of the 2014 IEEE International Conference on Consumer Electronics-Taiwan, Taipei, Taiwan, 26–28 May 2014; pp. 97–98. [PubMed] [CrossRef]

63. Burchardt, H.; Serafimovski, N.; Tsonev, D.; Videv, S.; Haas, H. VLC: Beyond point-to-point communication. *IEEE Commun. Mag.* **2014**, *52*, 98–105. [PubMed] [CrossRef]

64. Kumar, A.; Mihovska, A.; Kyriazakos, S.; Prasad, R. Visible light communications (VLC) for ambient assisted living. *Wirel. Pers. Commun.* **2014**, *78*, 1699–1717. [CrossRef] [CrossRef]

65. Varghese, A.; Tandur, D. Wireless requirements and challenges in Industry 4.0. In Proceedings of the 2014 International Conference on Contemporary Computing and Informatics (IC3I), Mysore, India, 27–29 November 2014; pp. 634–638. [PubMed] [CrossRef]

66. Cwikla, G.; Foit, K. Problems of integration of a manufacturing system with the business area of a company on the example of the Integrated Manufacturing Systems Laboratory. In Proceedings of the MATEC Web of Conferences, Brasov, Romania, 3–4 November 2016; Volume 94, p. 06004. [CrossRef] [CrossRef]

67. Cerruela Garcia, G.; Luque Ruiz, I.; Gomez-Nieto, M. State of the art, trends and future of bluetooth low energy, near field communication and visible light communication in the development of smart cities. *Sensors* **2016**, *16*, 1968. [CrossRef] [CrossRef]

68. Brena, R.F.; Garcia-Vazquez, J.P.; Galvan-Tejada, C.E.; Munoz-Rodriguez, D.; Vargas-Rosales, C.; Fangmeyer, J. Evolution of indoor positioning technologies: A survey. *J. Sens.* **2017**. [CrossRef] [CrossRef]

69. Memedi, A.; Tsai, H.M.; Dressler, F. Impact of realistic light radiation pattern on vehicular visible light communication. In Proceedings of the GLOBECOM 2017-2017 IEEE Global Communications Conference, Singapore, Singapore, 4–8 December 2017; pp. 1–6. [PubMed] [CrossRef]

70. Chen, L.; Wang, W.; Zhang, C. Coalition formation for interference management in visible light communication networks. *IEEE Trans. Veh. Technol.* **2017**, *66*, 7278–7285. [PubMed] [CrossRef]

71. Varanva, D.J.; Prasad, K.M. LED to LED communication with WDM concept for flash light of Mobile phones. *Edit. Preface* **2013**, *4*. [PubMed] [CrossRef]

72. Cui, Z.; Yue, P.; Ji, Y. Study of cooperative diversity scheme based on visible light communication in VANETs. In Proceedings of the 2016 International Conference on Computer, Information and Telecommunication Systems (CITS), Kunming, China, 6–8 July 2016; pp. 1–5. [PubMed] [CrossRef]

73. Mare, R.M.; Cugnasca, C.E.; Marte, C.L.; Gentile, G. Intelligent transport systems and visible light communication applications: An overview. In Proceedings of the 2016 IEEE 19th International Conference on Intelligent Transportation Systems (ITSC), Rio de Janeiro, Brazil, 1–4 November 2016; pp. 2101–2106. [PubMed] [CrossRef]

74. Alam, K.M.; Saini, M.; El Saddik, A. Toward social internet of vehicles: Concept, architecture, and applications. *IEEE Access.* **2015**, *3*, 343–357. [PubMed] [CrossRef]

75. Nitti, M.; Girau, R.; Floris, A.; Atzori, L. On adding the social dimension to the internet of vehicles: Friendship and middleware. In Proceedings of the 2014 IEEE international black sea conference on communications and networking (BlackSeaCom), Odessa, Ukraine, 27–30 May 2014; pp. 134–138. [PubMed] [CrossRef]

76. Maglaras, L.; Al-Bayatti, A.; He, Y.; Wagner, I.; Janicke, H. Social internet of vehicles for smart cities. *J. Sens. Actuator Netw.* **2016**, *5*, 3. [CrossRef] [CrossRef]

77. Yoo, J.H.; Jang, J.S.; Kwon, J. K.; Kim, H.C.; Song, D.W.; Jung, S.Y. Demonstration of vehicular visible light communication based on LED headlamp. *Int. J. Automot. Technol.,* **2016**, *17*, 347–352. [CrossRef] [CrossRef]

78. Uysal, M.; Ghassemlooy, Z.; Bekkali, A.; Kadri, A.; Menouar, H. Visible light communication for vehicular networking: Performance study of a V2V system using a measured headlamp beam pattern model. *IEEE Veh. Technol. Mag.* **2015**, *10*, 45–53. [CrossRef] [CrossRef]

79. Turan, B.; Narmanlioglu, O.; Ergen, S.C.; Uysal, M. Physical layer implementation of standard compliant vehicular VLC. In Proceedings of the 2016 IEEE 84th Vehicular Technology Conference (VTC-Fall), Montreal, Canada, 8–21 September 2016 ; pp. 1–5. [PubMed] [CrossRef]

80. Costanzo, A.; Loscri, V.; Costanzo, S. Software Defined Platforms for Visible Light Communication: State of Art and New Possibilities. *IEEE Commun. Soc. Multim. Commun. Tech. Committee* **2017**, *12*, 14–18. [PubMed]

81. Bhunia, S.; Sengupta, S. Implementation of interface agility for duplex dynamic spectrum access radio using USRP. In Proceedings of the MILCOM 2017–2017 IEEE Military Communications Conference (MILCOM), Maryland, USA, 23–25 October 2017; pp. 762–767. [PubMed] [CrossRef]

82. Baranda, J.; Henarejos, P.; Gavrincea, C.G. An SDR implementation of a visible light communication system based on the IEEE 802.15. 7 standard. In Proceedings of the ICT 2013, Litva, Czech Republic, 6–8 November 2013; pp. 1–5. [PubMed] [CrossRef]

83. Deng, P.; Kavehrad, M. Adaptive real-time software defined MIMO visible light communications using spatial multiplexing and spatial diversity. In Proceedings of the 2016 IEEE International Conference on Wireless for Space and Extreme Environments (WiSEE), Aachen, Germany, 26–29 September 2016; pp. 111–116. [PubMed] [CrossRef]

84. Deng, P. Real-Time Software-Defined Adaptive MIMO Visible Light Communications. In *Visible Light Communications*; IntechOpen: London, UK, 2017. [PubMed] [CrossRef]

85. Tsonev, D.; Chun, H.; Rajbhandari, S.; McKendry, J.J.; Videv, S.; Gu, E.; Haji, M.; Watson, S.; Kelly, A.E.; Faulkner, G.; et al. Dawson, M.D. A 3-Gb/s Single-LED OFDM-Based Wireless VLC Link Using a Gallium Nitride μLED. *IEEE Photonics Technol. Lett.* **2014**, *26*, 637–640. [PubMed] [CrossRef]

86. Bandara, K.; Niroopan, P.; Chung, Y.H. PAPR reduced OFDM visible light communication using exponential nonlinear companding. In Proceedings of the 2013 IEEE International Conference on Microwaves, Communications, Antennas and Electronic Systems (COMCAS 2013), Tel Aviv, Israel, 21–23 October 2013; pp. 1–5. [PubMed] [CrossRef]

87. Khalid, A.; Asif, H.M. NI cDAQ based software-defined radio for visible light communication system. In Proceedings of the 2017 2nd Workshop on Recent Trends in Telecommunications Research (RTTR), Palmerston North, New Zealand, 10 February 2017; pp. 1–5. [PubMed] [CrossRef]

88. Tsiropoulou, E.E.; Gialagkolidis, I.; Vamvakas, P.; Papavassiliou, S. Resource allocation in visible light communication networks: NOMA vs OFDMA transmission techniques. In Proceedings of the International Conference on Ad-Hoc Networks and Wireless, Lille, France, 4–6 July 2016; pp. 32–46. Springer, Cham, Switzerland. [CrossRef] [CrossRef]

89. Kizilirmak, R.C.; Rowell, C.R.; Uysal, M. Non-orthogonal multiple access (NOMA) for indoor visible light communications. In Proceedings of the 2015 4th International Workshop on Optical Wireless Communications (IWOW), Istanbul, Turkey, 7–8 September 2015; pp. 98–101. [PubMed] [CrossRef]

90. Yapici, Y.; Guvenc, I. Non-orthogonal multiple access for mobile VLC networks with random receiver orientation. *arXiv* **2018**, *arXiv:1801.04888*. [PubMed]

91. Saito, Y.; Kishiyama, Y.; Benjebbour, A.; Nakamura, T.; Li, A.; Higuchi, K. Non-orthogonal multiple access (NOMA) for cellular future radio access. In Proceedings of the 2013 IEEE 77th vehicular technology conference (VTC Spring), Dresden, Germany, 2–5 June 2013; pp. 1–5. [PubMed] [CrossRef]

92. Lin, B.; Ye, W.; Tang, X.; Ghassemlooy, Z. Experimental demonstration of bidirectional NOMA-OFDMA visible light communications. *Opt. Express.* **2017**, *25*, 4348–4355. [CrossRef] [CrossRef]

93. Tsiropoulou, E.E.; Vamvakas, P.; Papavassiliou, S. Resource Allocation in Multi-tier Femtocell and Visible-Light Heterogeneous Wireless Networks. *Resource Allocation in Next-Generation Broadband Wireless Access Networks*. IGI Global: Hershey, PA, USA, 2017. [PubMed] [CrossRef]

94. Singhal, C.; De, S. (Eds.) *Resource Allocation in Next-Generation Broadband Wireless Access Networks*; IGI Global: Hershey, PA, USA, 2017

95. National Instruments. Universal Software Radio Peripheral. NI USRP-2920/2921/2922 Datasheet. Available online: http://www.ni.com/pdf/manuals/376358a.pdf (access on 28 February 2019). [PubMed]

96. Mini-Circuits. Coaxial Bias-Tee. ZX85-12G-S+ Datasheet. Available online: https://www.minicircuits.com/pdfs/ZX85-12G-S+.pdf (access on: 28 February 2019). [PubMed]

97. Welch, T.B.; Shearman, S. Teaching software defined radio using the USRP and LabVIEW. In Proceedings of the 2012 IEEE International Conference on Acoustics, Speech and Signal Processing (ICASSP), Kyoto, Japan, 25–30 March 2012; pp. 2789–2792. [PubMed] [CrossRef]

98. Marriwala, N.; Sahu, O.P.; Vohra, A. LabVIEW based design implementation of M-PSK transceiver using multiple forward error correction coding technique for software defined radio applications. *J. Electrical Electron. Eng.* **2014**, *2*, 55–63. [PubMed] [CrossRef]

99. Haigh, P.A.; Burton, A.; Werfli, K.; Le Minh, H.; Bentley, E.; Chvojka, P.; Popoola, W.O.; Papakonstantinou, I.; Zvanovec, S. Zvanovec, S. A multi-CAP visible-light communications system with 4.85-b/s/Hz spectral efficiency. *IEEE J. Sel. Areas Commun.* **2015**, *33*, 1771–1779. [PubMed] [CrossRef]

100. Urick, V.J.; Qiu, J.X.; Bucholtz, F. Wide-band QAM-over-fiber using phase modulation and interferometric demodulation. *IEEE Photonics Technol. Lett.* **2004**, *16*, 2374–2376. [PubMed] [CrossRef]

101. Schmogrow, R.; Nebendahl, B.; Winter, M.; Josten, A.; Hillerkuss, D.; Koenig, S.; Meyer, J.; Dreschmann, M.; Huebner, M.; Koos,C.; et al Becker, J. Error vector magnitude as a performance measure for advanced modulation formats. *IEEE Photonics Technol. Lett.* **2012**, *24*, 61–63. [PubMed] [CrossRef]

102. Alonso, D.E. *Wireless Data Transmission for the Battery Management System of Electric and Hybrid Vehicles*; KIT Scientific Publishing: Karlsruhe, Germany, 2017; Volume 15. [PubMed]

103. Hamamatsu. Operates an APD with Single 5 v Supply (Standard Type, Short-Wavelength Type). Singh2018702 Datasheet. Available online: https://www.hamamatsu.com/resources/pdf/ssd/Singh2018702series_kacSingh201814e.pdf (accessed on 28 February 2019). [PubMed]

104. Hamamatsu. Detects Opcial Signals at 1 GHz, with High Sensitivity. Shen2017658 Datasheet. Available online: https://www.hamamatsu.com/resources/pdf/ssd/Shen2017658_kacLuo201423e.pdf (accessed on 28 February 2019). [PubMed]

105. Elamassie, M.; Karbalayghareh, M.; Miramirkhani, F.; Kizilirmak, R.C.; Uysal, M. Effect of fog and rain on the performance of vehicular visible light communications. In Proceedings of the 2018 IEEE 87th Vehicular Technology Conference (VTC Spring), Porto, Portugal, 3–6 June 2018; 1–6. [PubMed]

106. Hossain, F.; Afroze, Z. Eliminating the effect of fog attenuation on FSO link by multiple TX/RX system with travelling wave semiconductor optical amplifier. In Proceedings of the 2013 2nd International Conference on Advances in Electrical Engineering (ICAEE), Dhaka, Bangladesh, 19–21 December; pp. 267–272. [PubMed] [CrossRef]

107. Kim, I.I.; McArthur, B.; Korevaar, E.J. Comparison of laser beam propagation at 785 nm and 1550 nm in fog and haze for optical wireless communications. In Proceedings of the Optical Wireless Communications III, Boston, MA, USA, 6 February 2001; Volume 4214, pp. 26–38. [CrossRef] [CrossRef]

108. Ebrahim, K.J.; Al-Omary, A. Sandstorm Effect on Visible Light Communication. In Proceedings of the 2017 9th IEEE-GCC Conference and Exhibition (GCCCE), Manama, Bahrain, 8–11 May 2017; pp. 1–7. [PubMed] [CrossRef]

109. Luo, P.; Ghassemlooy, Z.; Le Minh, H.; Khalighi, A.; Zhang, X.; Zhang, M.; Yu, C. Experimental demonstration of an indoor visible light communication positioning system using dual-tone multi-frequency technique. In Proceedings of the 2014 3rd International Workshop in Optical Wireless Communications (IWOW), Funchal, Madeira Island, Portugal, 17 September 2014; pp. 55–59. [PubMed] [CrossRef]

110. Zhang, W.; Chowdhury, M.S.; Kavehrad, M. Asynchronous indoor positioning system based on visible light communications. *Opt. Engineering* **2014**, *53*, 045105. [CrossRef] [CrossRef]

111. Ganti, D.; Zhang, W.; Kavehrad, M. VLC-based indoor positioning system with tracking capability using Kalman and particle filters. In Proceedings of the 2014 IEEE International Conference on Consumer Electronics (ICCE), Las Vegas, NV, USA, 4–6 January 2014; pp. 476–477. [PubMed] [CrossRef]

112. Lin, B.; Tang, X.; Ghassemlooy, Z.; Lin, C.; Li, Y. Experimental demonstration of an indoor VLC positioning system based on OFDMA. *IEEE Photonics J.* **2017**, *9*, 1–9. [PubMed] [CrossRef]

113. Yamaguchi, S.; Mai, V.V.; Thang, T.C.; Pham, A.T. Design and performance evaluation of VLC indoor positioning system using optical orthogonal codes. In Proceedings of the 2014 IEEE Fifth International Conference on Communications and Electronics (ICCE), Danang, Vietnam, 30 July–1 August 2014; pp. 54–59. [PubMed] [CrossRef]

114. Huynh, P.; Yoo, M. VLC-based positioning system for an indoor environment using an image sensor and an accelerometer sensor. *Sensors* **2016**, *16*, 783. [CrossRef] [CrossRef]

115. Blinowski, G. Security issues in visible light communication systems. *IFAC-PapersOnLine* **2015**, *48*, 234–239. [CrossRef] [CrossRef]

116. Martinek, R.; Zidek, J. The real implementation of ANFIS channel equalizer on the system of software-defined radio. *IETE J. Res.* **2014**, *60*, 183–193. [CrossRef] [CrossRef]

117. Martinek, R.; Zidek, J. The real implementation of NLMS channel equalizer into the system of software defined radio. *Adv. Electr. Electron. Eng.* **2012**, *10*, 330–336. [PubMed] [CrossRef]

118. Martinek, R.; Konecny, J.; Koudelka, P.; Zidek, J.; Nazeran, H. Adaptive optimization of control parameters for feed-forward software defined equalization. *Wirel. Pers. Commun.* **2017**, *95*, 4001–4011. [CrossRef] [CrossRef]

119. Martinek, R.; Vanus, J.; Bilik, P.; Al-Wohaishi, M.; Zidek, J.; Wen, H. The implementation of equalization algorithms for real transmission channels. In Proceedings of the 2016 IEEE International Instrumentation and Measurement Technology Conference Proceedings, Taipei, Taiwan, 23–26 May 2016; pp. 1–6. [PubMed] [CrossRef]

120. Martinek, R.; Vanus, J.; Kelnar, M.; Bilik, P.; Zidek, J. Application of recursive least square algorithm to adaptive channel equalization. In Proceedings of the XXI IMEKO World Congress Measurement in Research and Industry, Prague, Czech Republic, 30 August–4 September 2015; IMEKO-International Measurement Federation Secretariat: Budapest, Hungary, 2015; pp. 1–4. [PubMed]

121. Martinek, R.; Razera, G.; Kahankova, R.; Zidek, J. Optimization of the training symbols for minimum mean square error equalizer. In Proceedings of the International Afro-European Conference for Industrial Advancement, Marrakesh, Morocco, 21–23 November 2016; Springer: Cham, Switzerland, 2016; pp. 272–287. [CrossRef] [CrossRef]

122. Yamazato, T.; Takai, I.; Okada, H.; Fujii, T.; Yendo, T.; Arai, S.; Andoh, M.; Harada, T.; Yasutomi, K.; Kagawa, K.; et al. Kawahito, S. Image-sensor-based visible light communication for automotive applications. *IEEE Commun. Mag.* **2014**, *52*, 88–97. [PubMed] [CrossRef]

123. Imai, Y.; Ebihara, T.; Mizutani, K.; Wakatsuki, N. Performance evaluation of high-speed visible light communication combining low-speed image sensor and polygon mirror in an outdoor environment. In Proceedings of the 2016 Eighth International Conference on Ubiquitous and Future Networks (ICUFN), Vienna, Austria, 5–8 July 2016; pp. 51–55. [PubMed] [CrossRef]

124. Yamazato, T.; Kawagita, N.; Okada, H.; Fujii, T.; Yendo, T.; Barai, S.; Kamakura, K. The uplink visible light communication beacon system for universal traffic management. *IEEE Access* **2017**, *5*, 22282–22290. [PubMed] [CrossRef]

electronics

MDPI

Article

Fingerprint-Based Indoor Positioning System Using Visible Light Communication—A Novel Method for Multipath Reflections

Huy Q. Tran and Cheolkeun Ha *

Robotics and Mechatronics Lab, University of Ulsan, Ulsan 44610, Korea; quanghuybkdn@gmail.com
* Correspondence: cheolkeun@gmail.com

Received: 30 November 2018; Accepted: 31 December 2018; Published: 6 January 2019

Abstract: A highly accurate indoor positioning under the effect of multipath reflections has been a prominent challenge for recent research. This paper proposes a novel indoor visible light communication (VLC) positioning model by connecting k-nearest neighbors (kNN) and random forest (RF) algorithms for reflective environments, namely, kNN-RF. In this fingerprint-based model, we first adopt kNN as a powerful solution to expand the number of input features for RF. Next, the importance rate of these features is ranked and the least effective one(s) may be removed to reduce the computation effort. Next, the training process using the RF algorithm is conducted. Finally, the estimation process is utilized to discover the final estimated position. Our simulation results show that this new approach improved the positioning accuracy, making it nearly five times better than other popular kNN algorithms.

Keywords: indoor positioning system (IPS); visible light communication (VLC); multipath reflections; k-nearest neighbors (kNN); random forest (RF)

1. Introduction

In the last few decades, the global positioning system (GPS) has been widely employed in positioning and navigation because of its high reliability and accuracy, and because of its real-time positioning capability [1]. In indoor environments, however, this power substantially declines and it may even be impossible to accurately and continuously locate the coordinates of the followed object, largely due to building structures, building materials, and other obstructions [2].

Recently, with the rapid development of high-intensity, solid-state white LEDs, the idea of using LEDs for both illumination and data transmission has become a promising trend in contemporary applications [3], particularly the growing demand for positioning techniques in indoor environments where people spend approximately 90% of their time during the day [4]. Also, LED lighting technology is gradually replacing conventional indoor lighting methods due to its safety, economy, and environmental friendliness [3,5]. Additionally, visible light communication (VLC) has recently emerged as one of the most promising candidates for fifth-generation (5G) mobile communications [1]. Several other types of wireless signals have been applied in indoor positioning fields, such as WiFi, RFID, Bluetooth, and Zigbee, and each of them has unique strengths, although their positioning accuracy remains somewhat limited [2].

In this paper, we propose a novel kNN-RF model based upon the traditional fingerprint technique. Unlike the existing received signal strengths (RSS)-based algorithms, in this scheme k-nearest neighbors (kNN) is responsible for enlarging the number of inputs for the next random forest (RF) model, which needs enough data to avoid overfitting and to obtain more accurate training results. Not all data generated from kNN are used by the RF model; rather, the data are sent through an evaluation and collection process which ranks the importance of each feature and eliminates the useless ones.

This process reduces the computational time and increases the capacity of the RF training operation. Another noteworthy point is that our proposed model incorporates the strong effects of multipath reflection, ambient light, and other noises. Our approach produced a remarkable reduction in the error range from 94.2 cm by the conventional kNN method to 19.3 cm by the kNN-RF method.

The main contributions of this paper are as follows:

- Most of recent VLC-based research have ignored the impact of reflection (see Section 2). However, this noise exerts a very bad influence on the positioning accuracy, especially in areas outside the room's center. In this paper, the effects of multipath reflection are thoroughly explored at the highest reflective rate.
- We propose a novel machine learning-based indoor positioning solution. This method first uses kNN as a powerful tool to expand the number of features for RF in the next step. Then the position of the mobile object can be located based on the estimation process in the RF algorithm which collects all the features directly from both the LED lights and the kNN output signals.

The remainder of this paper is organized as follows. Section 2 summarizes and analyzes some related works. Section 3 introduces the system model with and without multipath reflections, and then presents our proposed algorithm in detail. Section 4 discusses our simulation results and performance evaluations, then some proposed real applications are presented in Section 5, and Section 6 offers our discussion and conclusion.

2. Related Work

To improve accuracy, several VLC-based positioning techniques have been proposed [6], including the angle of arrival (AOA) [7], time of arrival (TOA), time difference of arrival (TDOA) [8,9], phase difference of arrival (PDOA) [10,11], and RSS [12,13]. Each of them has distinct limitations: AOA helps considerably to reduce the positioning errors but requires high computational complexity; TOA and TDOA need to ensure compliance with strict requirements for synchronization; PDOA requires a lower computational time than TDOA but needs a local oscillator; and RSS-based positioning accuracy is considered moderately low because of the effect of ambient light, noises, and the tilt angle of the photo-detector (PD) [10].

In addition to diverse types of signal-based positioning methods, a few positioning algorithms have been developed to optimize the quality of indoor positioning systems (IPS). In Reference [12], the authors proposed an improved kNN model by applying weights to more accurately determine the Euclidean distances.

Another approach, which compares machine learning and other conventional RSS-based solutions, was presented in Reference [13]. However, the general focus of these articles ignores multipath reflections, even though they actually exist and have an enormous impact on the system quality. To elucidate these impacts, the authors in Reference [14] proved that the influence of multipath reflections is much weaker and seems to be almost unaffected in the central region of a given room. In contrast, the reflection in other areas (i.e., the corners and edges) is more powerful than in the case of no reflection. In the given results, the positioning errors in the corner unexpectedly increased and reached roughly 2 m.

In Reference [15] the authors used the trilateration method to show that the positioning errors increased by 76 times, from 2 cm using line-of-sight (LOS) compared to 152 cm with non-line-of sight (NLOS). In Reference [16], the authors proposed a new approach to reduce the positioning error by using calibration methods, including selecting the strongest LED signal and decreasing the distance between the LED bulbs. As a result, there was a considerable improvement in the positioning accuracy, and the errors in the whole room were at 0.3 m and 0.25 m as a result of LED selection and changing the LED distance, respectively. However, the corresponding errors outside the central region were still high, at 0.47 m and 0.36 m. Additionally, a change in the number of LEDs and the distance between them may lead to changes in the structure of the room and illumination intensity. These analyses demonstrate that interference exerts a very negative influence on the positioning accuracy of LOS and NLOS, especially in corners and near edge areas. To reduce such detrimental effects, the authors of Reference [17] presented a new way to improve the positioning accuracy in a room corner, based on tilting the image sensor. They did not consider the multipath reflections and still achieved low accuracy at some points, and found it difficult to determine the optimal tilt angle.

3. System Model and Proposed Positioning Method

In this section, we first investigate the system model used for the proposed solution. This model consists of a directed optical channel and a non-directed optical channel. In each part, we focus on calculating the received optical power, then the total received optical power from both channels are shown. Next, a typical system configuration and our proposed kNN-RF algorithm are introduced in detail.

3.1. System Model

3.1.1. Directed Optical Channel

The directed optical channel, or LOS path, is shown in Figure 1. This channel is of vital importance in almost every indoor VLC positioning system because it receives most of the total optical power. As illustrated in Figure 2, the maximum received power is 0.92 mW and the minimum is 0.34 mW. The power distribution in the central region appears quite uniform, and the illumination intensity in this area is stronger than other areas, especially in the corners.

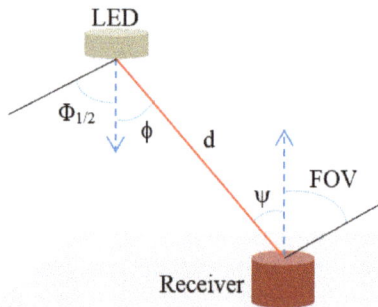

Figure 1. Directed channel model with one LED.

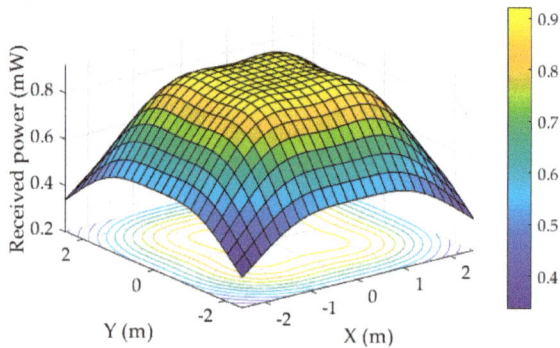

Figure 2. Received power distribution with directed channel.

To compute the LOS power, we assume that the distance and the angle from the transmitter to the receiver are d and φ, respectively. The irradiance angle φ and incidence angle Ψ are of equal value because we accept that both the LED and the photosensitive area of the receiver are parallel to the floor (see Figure 1).

To effectively detect the presence of visible light from the LED, the signal collection area on the optical sensor plays an important role. This area has a great influence over the positioning correctness and can be calculated as [18]:

$$A_{esc} = \begin{cases} A_{PD} T_s(\Psi) g(\Psi) \cos(\Psi), & 0 \le \Psi \le \Psi_c, \\ 0, & \Psi > \Psi_c \end{cases} \tag{1}$$

where A_{PD} is the active detector area of the PD; $T_s(\Psi)$ is the gain of the optical filter; $g(\Psi)$ is the gain of the optical concentrator; and Ψ_c is the receiver field of view (FOV).

The irradiance (W/cm^2) is given by Reference [18]:

$$I_s(d, \phi) = \frac{P_t R_o(\phi)}{d^2} \tag{2}$$

where P_t is the transmitted power; d is the transmission distance from the LED to the PD; and $R_o(\phi)$ is the Lambertian radiant intensity, written as [19]:

$$R_o(\phi) = \left[\frac{n+1}{2\pi}\right] \cos^n(\phi) \tag{3}$$

where n is the Lambertian order and is determined from [19]:

$$\Phi_{1/2} = \cos^{-1}\left(\frac{1}{2}\right)^{\frac{1}{n}} \tag{4}$$

Hence,

$$n = \frac{-\ln(2)}{\ln(\cos \Phi_{1/2})} \tag{5}$$

The optical power gathered from the LED lamps can be expressed as [18]:

$$P_{LOS} = I_s(d, \phi) A_{esc} \tag{6}$$

Finally, the total directed optical power at the receiver is as follows [5]:

$$P_{LOS} = \begin{cases} P_t \frac{(n+1)A_{PD}}{2\pi d^2} \cos^n(\phi) T_s(\Psi) g(\Psi) \cos(\Psi), & 0 \le \Psi \le \Psi_c, \\ 0, & \Psi > \Psi_c \end{cases} \tag{7}$$

3.1.2. Non-Directed Optical Channel

To evaluate the impact of multipath noises on the system performance, we conducted a comprehensive study of the effects of the first reflection from the four walls around the room. In practice, the reflection of visible light always exists in an indoor environment, especially in a narrow space with walls, ceiling, floor, and some pieces of furniture. As can be seen in Figure 3, the active area of the receiver is facing upwards and is parallel with the floor. Hence, it is impossible to have a sense of the first signal reflected from the floor. In addition, the direction of the LED bulb is facing downwards, and we also set the half-power semi-angle of each bulb at 60° and the distance from the light source to the ceiling at 0.7 m. This meant that the amount of light which the ceiling directly received from the LED bulb was completely restricted. Furthermore, any additional reflections could be neglected because they would have no significant effect compared to the total noise level power [20]. For these reasons, we chose the first reflection from the walls and the LOS signal as the major channel model in our work. Figure 4 illustrates that the power distribution in the corner and near the wall was higher than in the central area.

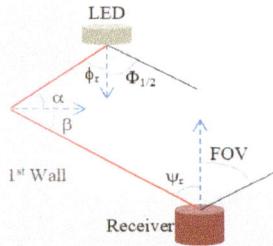

Figure 3. Non-directed channel model with one LED.

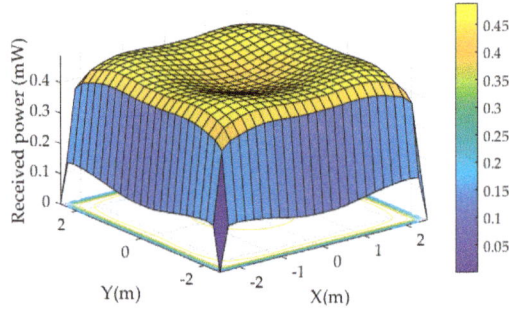

Figure 4. Received power distribution with non-directed channel.

To compute the received power of the first reflection link, we compute the received power at a reflective point on the wall as follows:

$$P_{R\_wall} = I_s(d_1, \phi) A_{eff\_wall} \tag{8}$$

where $I_s(d_1, \phi)$ can be obtained from Equation (2); d_1 is the distance between an LED and a reflective point; and A_{eff_wall} is the effective signal collection area on the wall and is given as follows:

$$A_{eff\_wall} = dA_{wall} \cos(\alpha) \tag{9}$$

where dA_{wall} is the surface element on the wall and α is the angle of irradiance from an LED to the reflective point.

Based on the cosine law and the inverse square law, the transmitted power from a reflective point is then given:

$$P_{T\_wall} = \frac{P_{r\_wall}\rho}{d_2^2} \cos(\beta) \tag{10}$$

where d_2, β are the distance and the angle, respectively, from a reflective point and the receiver, and ρ is the reflectance factor.

Substituting Equations (8) and (9) into (10), we have:

$$P_{T\_wall} = P_t \frac{m+1}{2\pi d_1^2 d_2^2} \rho \cos^n(\phi) dA_{wall} \cos(\alpha) \cos(\beta) \tag{11}$$

From here, the PD uses P_{T_wall} as the input signal from the non-direct link, and the final power P_{Diff} that the sensors collect can be calculated as:

$$P_{Diff} = P_{T\_wall} A \cos(\Psi_r) \, Ts(\Psi_r) g(\Psi_r) \tag{12}$$

where Ψ_r is the incidence angle of the light from the wall.

Finally, we adopt the non-directed optical power as follows [5]:

$$P_{Diff} = P_t \frac{A(n+1)}{2\pi d_1^2 d_2^2} \rho \cos^n(\phi) dA_{wall} \cos(\alpha) \cos(\beta) \cos(\Psi_r) Ts(\Psi_r) g(\Psi_r), \; 0 \le \Psi_r \le \Psi_c \tag{13}$$

3.1.3. Overall Optical Channel

In this paper, the overall channel is considered as the connection between the LOS path and the diffuse reflection path. The purpose of this combination was to further investigate the undesirable effects of reflection on the system performance and to make our simulation results more reliable and practical. Furthermore, as previously discussed in Sections 3.1.1 and 3.1.2, we simply focused on the first reflection path from the four walls around the empty room and thus, the total received optical power from the four LED groups is the sum of the directed and non-directed optical power (see Figure 5) given by:

$$P_r = P_{Los} + P_{Diff} \tag{14}$$

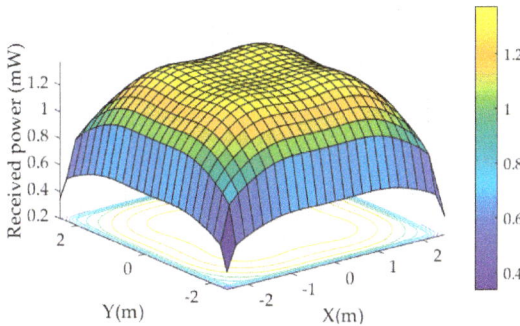

Figure 5. Received power distribution with overall channel.

3.2. Proposed Positioning Method

3.2.1. System Configuration

An overview of a typical room model is shown in Figure 6 [14], where its length, width, and height are fixed at 5 m, 5 m, and 3 m, respectively. We assume the LED bulbs are installed on an imaginary plane at a height of 2.3 m from the floor. Four LED bulbs are used in this model, and each of them has 10 W of transmitted optical power and a half-power semi-angle of exactly 60°. In our system, the receiver moving around the 25-m^2 floor has a PD active area of 1 cm^2 and is parallel with the floor. Other technical parameters related to the transmitters, receiver, and multipath reflection noise are summarized in detail in Table 1.

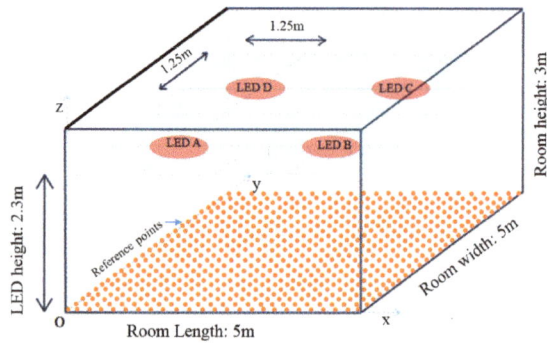

Figure 6. System configuration.

Table 1. Model technical parameters.

Parameter	Value
Room dimension (L × W × H)	5m × 5m × 3m
Reflection coefficient	0.8
Transmitters	
LED power per group	10 W
No. of LED groups	4
Wave length	420 nm
Elevation:	$-90°$
LED position (x, y, z) (m)	A (1.25, 1.25, 2.3)
	B (3.75, 1.25, 2.3)
	C (3.75, 3.75, 2.3)
	D (1.25, 3.75, 2.3)
Half power semi-angle	60°
Receiver	
PD active area	1 cm^2
FOV	60°
Gain of optical filter	1
Receiver sensitivity	-30 dBm
$T_s(\Psi)$	1
$g(\Psi)$	1.5

3.2.2. kNN-RF Algorithm

In the following section, we present a novel fingerprint location method known as kNN-RF, a combination of kNN and RF. In particular, the training data are expanded by kNN, and this leads to a significant improvement in the positioning accuracy after the RF training process.

The proposed RSS-based technique can be divided into two main phases— the offline phase and the online phase (Figure 7).

The offline phase plays a key role since it takes most of the program execution time. This phase consists of three separable basic steps. First, based on the area of the floor (see Figure 6), we prepared a feasible number of reference points (25 × 25) with the distance between these points of 20.833 cm and the optical receiver on a mobile object then began to collect all the RSS data from the four LED bulbs at each predetermined point. Second, all the sampling data gathered from the previous step were stored in memory.

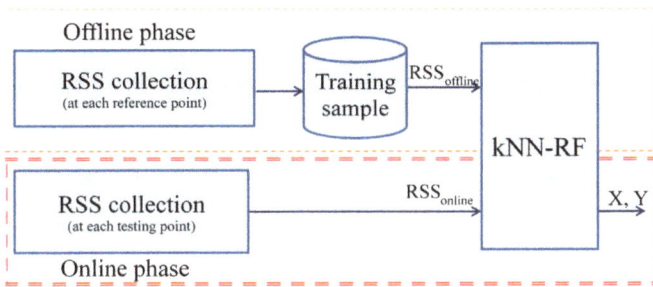

Figure 7. Block diagram of the proposed system.

Finally, all these data were sent to the kNN-RF model. This process (Figure 8) is described in the following details:

(i) Feature expanding process using kNN: First using kNN is an important step to increase the number of variables for the RF model, by finding the closest points to each reference point. kNN is a very simple machine learning algorithm and is widely used in the current IPSs. The main objective of this method is to find the Euclidean distances based on all of the reference points collected in the offline phase. For the number of Euclidean distances, it is entirely possible to rely on the value of *k* (*k* > 1) [12]. Specifically, the Euclidean distances are computed as follows:

$$DE = \sqrt{\sum_{i=1}^{4} \left(RSS_{on\_i} - RSS_{off\_i} \right)^2} \tag{15}$$

where *i* is the number of LED bulbs (i.e., *i* = 4); RSS_{on_i} are the RSS values collected in the online phase; and RSS_{off_i} are the RSS values stored in the offline phase.

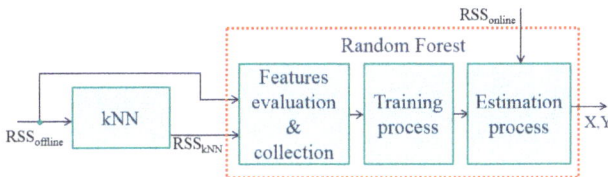

Figure 8. kNN-RF model.

After determining the Euclidean distances, the coordinates of each closest point are easily identified. From this location, their RSS values are also inferred. To determine one of the two inputs of the next RF block, the major operation of this step is to identify the nearest RSS of each collected fingerprint based on the Euclidean distance (15). The number of RSS outputs from kNN always depends on how many *k* closest points are used. In our proposed system, the number of LED bulbs is fixed at four; hence each *k* value corresponds to four RSS values. In summary, RSS data transferred to the next step from the kNN model are illustrated in Table 2, with *k* = 2 and given as:

$$RSS_{kNN} = LED_{no} k F_{off} \tag{16}$$

where RSS_{kNN} is the total number of RSS values after processing using kNN; LED_{no} is the number of LEDs (here, LED_{no} = 4); *k* is the number of nearest points; and F_{off} is the number of fingerprints (here, F_{off} = 625).

(ii) Feature evaluation and collection process: The RSS signals from both the kNN output and the sampling data stored in the memory are transferred to the first function block in the RF model, namely, feature evaluation and collection. As discussed earlier, the number of features from kNN varies depending upon the number of *k*-nearest points, and they obviously exist in direct proportion. In the RF algorithm we can improve the performance of the training process and reduce the computation cost using the feature importance ranking [21,22].

Table 2. Received optical power using kNN algorithm with two nearest points (Unit: W).

	RSS (online)				RSS ($k = 1$)				RSS ($k = 2$)			
No.	LED A	LED B	LED C	LED D	LED A	LED B	LED C	LED D	LEDA	LED B	LED C	LED D
1	7.13×10^{-5}	1.16×10^{-5}	1.16×10^{-5}	4.52×10^{-6}	7.13×10^{-5}	1.16×10^{-5}	1.16×10^{-5}	4.52×10^{-6}	10.3×10^{-5}	1.64×10^{-5}	2.34×10^{-5}	9.24×10^{-5}
2	10.3×10^{-5}	1.64×10^{-5}	2.34×10^{-5}	9.24×10^{-6}	10.3×10^{-5}	1.64×10^{-5}	2.34×10^{-5}	9.24×10^{-6}	10.3×10^{-5}	2.34×10^{-5}	1.64×10^{-5}	9.24×10^{-6}
3	11.4×10^{-5}	1.70×10^{-5}	2.41×10^{-5}	9.02×10^{-6}	11.4×10^{-5}	1.70×10^{-5}	2.41×10^{-5}	9.02×10^{-6}	12.2×10^{-5}	1.73×10^{-5}	2.54×10^{-5}	8.92×10^{-6}
4	12.2×10^{-5}	1.73×10^{-5}	2.54×10^{-5}	8.92×10^{-6}	12.2×10^{-5}	1.73×10^{-5}	2.54×10^{-5}	8.92×10^{-6}	12.7×10^{-5}	1.75×10^{-5}	2.78×10^{-5}	9.02×10^{-6}
5	12.7×10^{-5}	1.75×10^{-5}	2.78×10^{-5}	9.02×10^{-6}	12.7×10^{-5}	1.75×10^{-5}	2.78×10^{-5}	9.02×10^{-6}	13.0×10^{-5}	1.75×10^{-5}	3.12×10^{-5}	9.30×10^{-6}
...	...	...	...	...	...	...	...	...	...	...	...	...
621	9.02×10^{-6}	2.78×10^{-5}	1.75×10^{-5}	12.7×10^{-5}	9.02×10^{-6}	2.78×10^{-5}	1.75×10^{-5}	12.7×10^{-5}	9.30×10^{-6}	3.12×10^{-5}	1.75×10^{-5}	13.0×10^{-5}
622	8.92×10^{-6}	2.54×10^{-5}	1.73×10^{-5}	12.2×10^{-5}	8.92×10^{-6}	2.54×10^{-5}	1.73×10^{-5}	12.2×10^{-5}	9.02×10^{-6}	2.78×10^{-5}	1.75×10^{-5}	12.7×10^{-5}
623	9.02×10^{-6}	2.41×10^{-5}	1.70×10^{-5}	11.4×10^{-5}	9.02×10^{-6}	2.41×10^{-5}	1.70×10^{-5}	11.4×10^{-5}	8.92×10^{-6}	2.54×10^{-5}	1.73×10^{-5}	12.2×10^{-5}
624	9.24×10^{-6}	2.34×10^{-5}	1.64×10^{-5}	10.3×10^{-5}	9.24×10^{-6}	2.34×10^{-5}	1.64×10^{-5}	10.3×10^{-5}	9.24×10^{-6}	1.64×10^{-5}	2.34×10^{-5}	10.3×10^{-5}
625	4.52×10^{-6}	1.16×10^{-5}	1.16×10^{-5}	7.13×10^{-5}	4.52×10^{-6}	1.16×10^{-5}	1.16×10^{-5}	7.13×10^{-5}	9.24×10^{-6}	1.64×10^{-5}	2.34×10^{-5}	10.3×10^{-5}

In Figure 9 the importance level of the variables is illustrated, with *k* = 2, 3, 4, and 5. In each case there are always differences among the feature groups (e.g., RSS_online, RSS_*k*1, RSS_*k*2, RSS_*k*3, RSS_*k*4, RSS_*k*5), and even among elements in a group. The least important value can be neglected to achieve better or at least the same positioning accuracy using much less computing time.

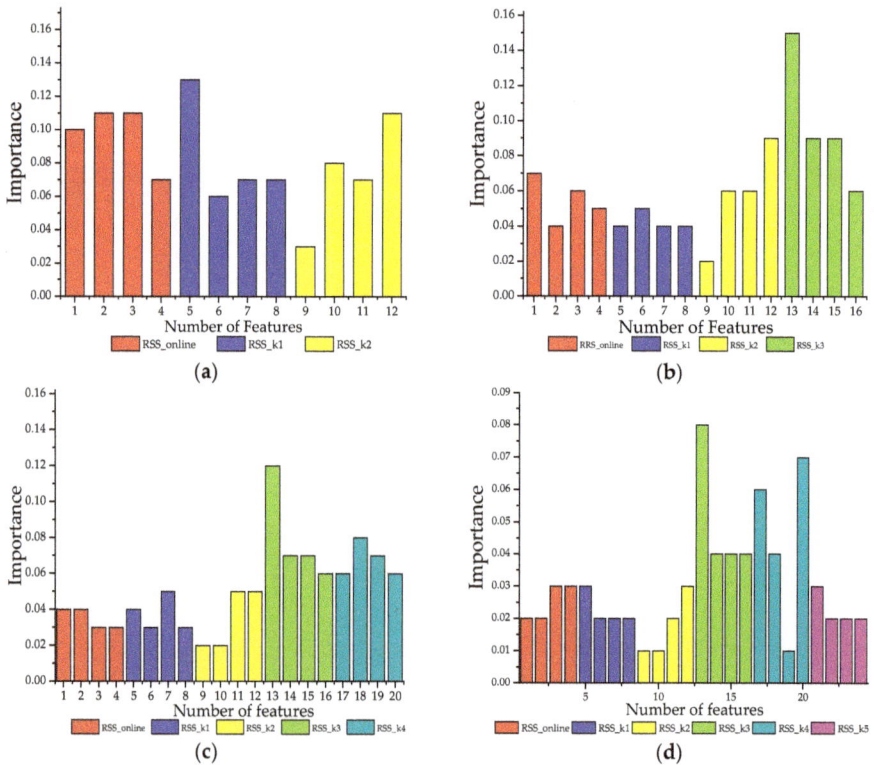

Figure 9. Feature importance ranking by (**a**) *k* = 2, (**b**) *k* = 3, (**c**) *k* = 4, (**d**) *k* = 5.

(iii)   The training process using RF: After evaluating and collecting the notable features, the RF training process is executed. We set the number of trees and features to 40 and 12, respectively. Similar to the previous procedure in kNN, the RF also uses the RSS data as the input signals, but the difference here is the combination of RSS data collected from the reference points and the RSS data that came from the kNN. This connection increases the number of variables, which helps reduce the chances of overfitting. However, a tremendous amount of data can lead to a very long run time.

Thus, the analysis results in Figure 9 can be used to choose, as well as remove, the appropriate features. Then, as shown in Figure 10, by neglecting some features that have an importance rate ≤4%, the CPU time is achievably enhanced, by 5.99% with $k = 2$ and by 18.3% with $k = 3$. With $k = 5$, half of all the features are skipped and the improvement increases up to 31%, while the mean errors rise only slightly by 7.45%. Therefore, the CPU time gradually decreases as more useless features are neglected.

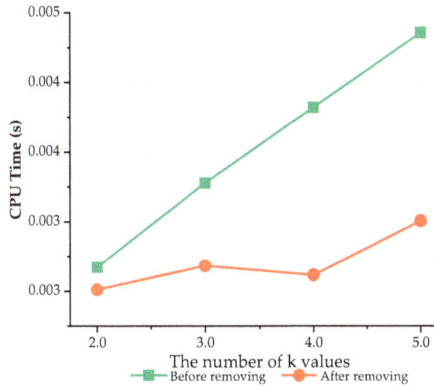

Figure 10. CPU time comparison before and after collection process.

In the online phase, the current RSS data from all the LED lights are sent directly to the estimation process block in the kNN-RF model (Figure 8). This block then uses these data and all the trained data in the offline phase to estimate the position of the object being considered. Estimation plays a major role in the online phase and has an effect on the final estimated position.

To evaluate the performance of our proposed algorithm, we shuffled the dataset randomly and divided them into 10 subsets. We then performed cross validations in each of them. By using this method, each subset is given the opportunity to be a testing set 1 time and be a training set 9 times. The results shown in Figure 11 demonstrate that our method is able to predict unseen data with a mean accuracy of 99.33%. More in-depth analyses of the impacts of the number of neighbors in kNN, the number of trees in RF, the receiver angle, and the performance of our method are thoroughly discussed in the next section.

10-fold	DATASET										Accuracy (%)
Fold 1	Test	Train	Train	Train	Train	Train	Train	Train	Train	Train	99.48
Fold 2	Train	Test	Train	Train	Train	Train	Train	Train	Train	Train	99.55
Fold 3	Train	Train	Test	Train	Train	Train	Train	Train	Train	Train	99.60
Fold 4	Train	Train	Train	Test	Train	Train	Train	Train	Train	Train	99.15
Fold 5	Train	Train	Train	Train	Test	Train	Train	Train	Train	Train	99.32
Fold 6	Train	Train	Train	Train	Train	Test	Train	Train	Train	Train	99.16
Fold 7	Train	Train	Train	Train	Train	Train	Test	Train	Train	Train	99.39
Fold 8	Train	Train	Train	Train	Train	Train	Train	Test	Train	Train	99.09
Fold 9	Train	Train	Train	Train	Train	Train	Train	Train	Test	Train	99.28
Fold 10	Train	Train	Train	Train	Train	Train	Train	Train	Train	Test	99.29

Figure 11. Cross validations.

4. Performance Evaluation

4.1. Effects of the Number of Neighbors in kNN

We examined cases with various numbers of k nearest points, and from Table 3 it is virtually certain that when k gradually increased from 2 to 8, the accuracy score showed only a minor fluctuation. However, some error types, such as the mean absolute error (MAE), mean square error (MSE), and root mean square error (RMSE) were slightly worse, corresponding to the increase in k values. As mentioned in Section 3.2, finding an ideal k value is considered an excellent way to satisfy the real-time requirements and enhance the quality of the recommended solution. In our proposed method, the number of features sent to the RF block increased four times, corresponding to each k value added (see Section 3.2). Therefore, in this paper, k is set to 2.

Table 3. Positioning accuracy vs. the number of k.

The Number of k	2	3	4	5	6	7	8
Accuracy Score (%)	99.08	99.34	99.06	99.10	99.18	99.27	98.98
MAE (m)	0.132	0.146	0.150	0.151	0.156	0.159	0.163
MSE (m)	0.039	0.040	0.044	0.044	0.047	0.048	0.051
RMSE (m)	0.196	0.200	0.208	0.209	0.216	0.217	0.225

4.2. Effects of the Number of Trees in RF

The number of trees in RF has a significant impact on not only the computational time, but also the positioning error of the system. To achieve reliable results, we randomly selected 125 testing data points (20% overall) and executed 100 consecutive repetitions to evaluate the influence of the number of trees on the system performance.

As depicted in Figure 12, the more trees that are chosen, the smaller the mean of the positioning errors. When the number of trees is more than 40, the performance continues to slightly improve, but not significantly. With 40 trees, the mean of the positioning errors was 0.193 m, and with 150 trees it was 0.189 m, a difference of only about 2.07%. From 70 trees onwards, the system became more stable, with mean errors in the range of 0.189 m to 0.191 m.

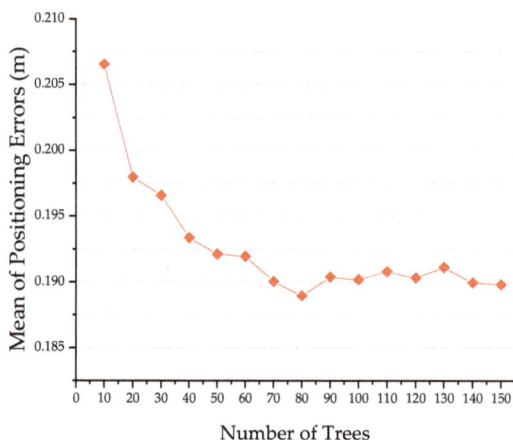

Figure 12. The number of trees vs. mean errors.

We also found that there were no considerable differences in the mean errors when the number of trees reached 40 and more. However, as discussed in Section 3.2, the tree numbers are directly proportional to the computational effort required (Figure 13). For instance, if 150 trees are selected instead of 10, the computational time will increase more than 13 times; thus, our system always suffers from a heavy computation burden. Therefore, to ensure the real-time performance and acceptable positioning quality, we recommend limiting the number of potential trees to the range of 40 to 70; even smaller values can be considered when strict execution time is required.

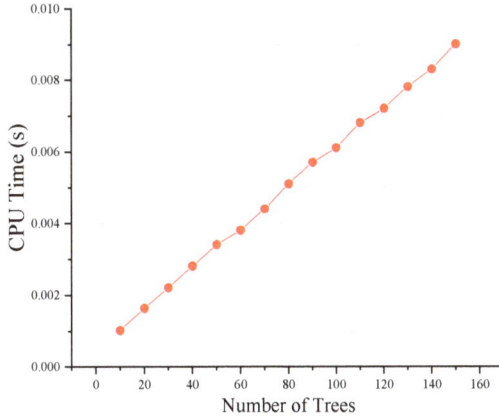

Figure 13. The numbers of trees vs. CPU time.

4.3. Effects of the Receiver Angle

In practice, the angle of the receiver may change over time and the degree of change depends on either the robustness of the mobile object equipped with a PD, the flatness of the moving surface, or the posture of the person using the PD sensor. In this section, we evaluated the effects of the receiver angle on the positioning accuracy by testing the mean of positioning errors corresponding to three separate cases: $10°$, $20°$, and $30°$. For simplicity we assumed that the receiver angle is formed by the horizontal plane of the PD and the horizontal plane of the floor. This can be seen in Figure 14, compared with the case of a fixed rotation angle ($0°$), the positioning errors in the cases of $10°$, $20°$, and $30°$ inclinations increased 3%, 28%, and 34%, respectively. This means that as we gradually increase the tilt angle of the optical receiver, the positioning accuracy changes in a deteriorating direction. However, the positioning error showed a slight increase in the range from a $0°$ to $10°$ inclination of only 3%. This range of tilt angle is also one of the most ubiquitous situations in reality. We are, therefore, confident that our proposed solution fully satisfies the actual requirements.

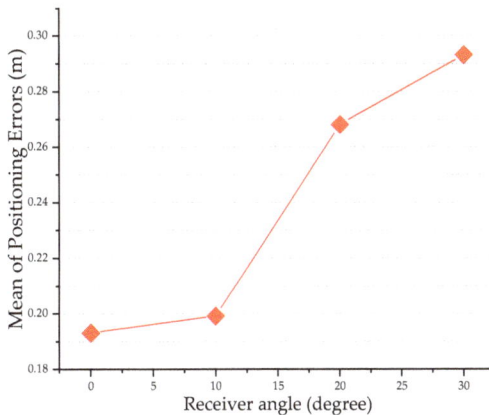

Figure 14. Positioning accuracy in different receiver angles.

4.4. Performance Evaluation

In Figure 15, the true positions (tails) and the estimated positions (heads) are used to visually illustrate the positioning accuracy of trilateration, kNN, Weighted-kNN (WkNN), and kNN-RF algorithms. It is clear that the combination of kNN and RF produced the most promising performance.

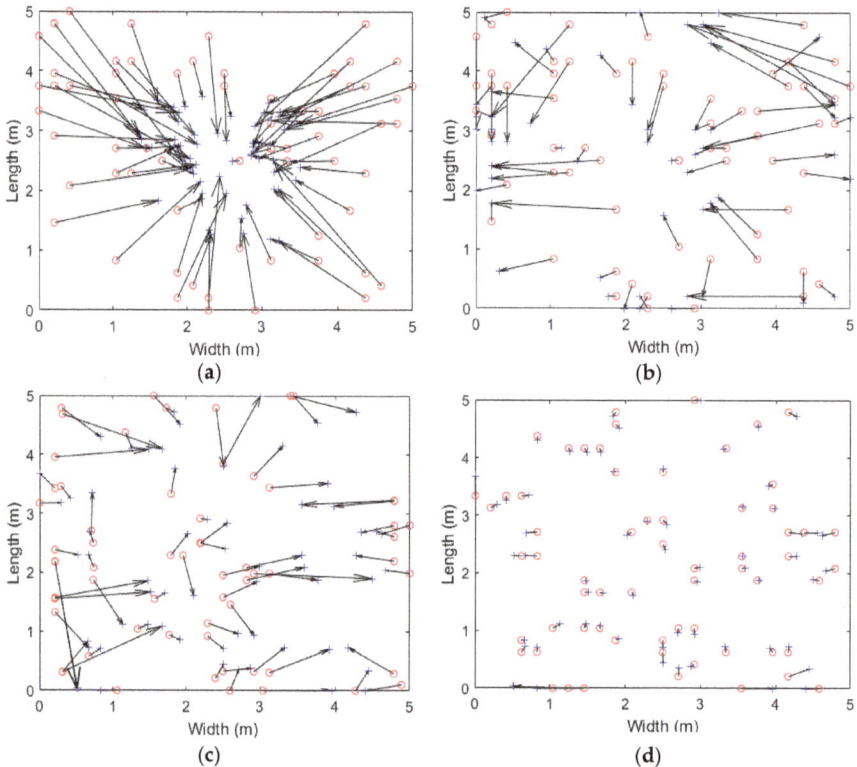

Figure 15. Position estimation outcome by (a) Trilateration, (b) kNN, (c) WkNN, and (d) kNN-RF.

In Reference [12], the authors demonstrated that a WkNN model in a non-reflective space was approximately 36% to 50% more accurate than the trilateration method with and without ambient light, respectively. In a reflective environment, however, the precision of the positioning system with the WkNN method became much worse, as the mean error rose from 0.031 m to 0.92 m, a 29-fold increase. In contrast to the first three algorithms (i.e., trilateration, kNN, and WkNN), the errors in the kNN-RF method were considerably lower, with an RMSE of 0.193 m (Figure 16).

As shown in Figure 17, the kNN-RF showed positive effects in specific areas of the model. By analyzing the random data with kNN and kNN-RF, we found that the highest error appeared in the corner where the impacts of noise and reflection were greatest. However, after being trained by the kNN-RF algorithm, the error dramatically declined (almost 7-fold) from 1.16 m to 0.166 m. Although the effect of the reflection in the region near the wall was higher than in the central area, and was lower than in the corner area, there was a notable (6-fold) mean error improvement. In contrast, due to the limited impact of the reflection on the central points, that area achieved the best positioning accuracy, 0.069 m, and thus gained only a slight improvement from kNN to kNN-RF.

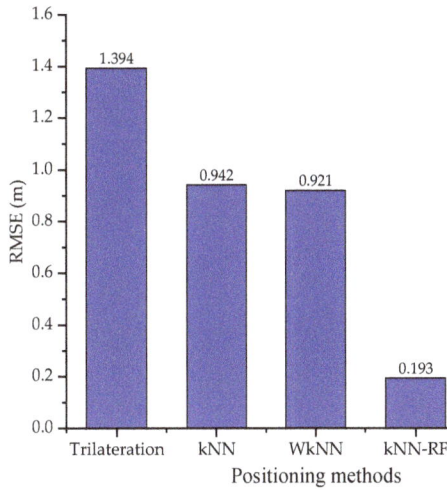

Figure 16. Performance comparison.

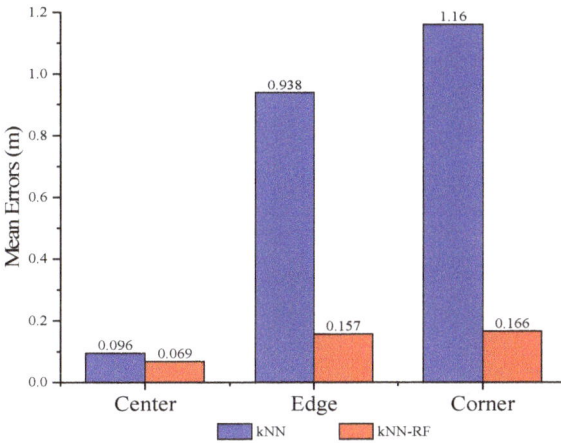

Figure 17. Positioning accuracy in different places.

These results demonstrate that our proposed kNN-RF solution produced a significant improvement in the accuracy of the positioning system, even though the interference intensity became stronger due to reflection and other noises.

5. Applications

The evolution of indoor positioning system using LED light gives us an ideal opportunity to develop useful products that serve our daily lives. In particular, we can build in-house localization applications at a low cost thanks to the availability of LED bulbs. In this paper, we suggest some applications that we are pursuing.

Assistive devices for people with disabilities: The PD will be put on a smart cane or other assistive walking devices to localize the position of the blind users and help them reach their destination quickly and safely. Additionally, a smart wheel chair with a positioning function enables the elderly and the disabled to travel easily. This can help improve their quality of life and maintain their sense of independence.

Indoor mobile robots: In addition to the applications for the disabled, LED-based indoor positioning solution is also particularly useful in the field of mobile robots. We can use a mobile robot as an effective means of transport in a factory or warehouse. The robot is used to pick up, move, and place objects automatically. We can also adopt this kind of robot as a home assistant device which can make our lives easier and more comfortable.

6. Discussion and Conclusions

To improve the positioning accuracy in multipath reflective environments, an innovative indoor VLC positioning model is proposed in this paper, namely kNN-RF. Our simulation results show that this combination of kNN and RF algorithms was five times more precise than other kNN methods. As for conventional kNN methods, the quality of the positioning system is satisfactory if and only if there is no existence of reflection. This noise, however, always persists in the indoor environment. In this paper, we consider not only the highest reflection rate, but also the effect of ambient light, thermal noise, and shot noise. The obtained outcomes show a considerable improvement in the positioning accuracy outside of the room center, which faces the highest multipath reflections. For instance, the positioning errors represent a marked improvement from kNN to kNN-RF in the room corners and near the wall by 7-fold and 6-fold, respectively.

The potential results of the kNN-RF method present a chance of improving the performance of the positioning system despite the negative impact of noises.

To further our research, we plan to conduct a real-time, full-scale experiment to demonstrate the performance of our proposed method. We will also focus on improving the positioning accuracy by strengthening the optimization of signal pre-processing before taking the kNN-RF algorithm into account. After performing some improvements to the algorithm, we will try to apply our proposed algorithm to indoor a mobile robot. For our current study, a VLC-based positioning system is one of the main parts in the field of a multi-robot system, including collision avoidance and multi-robot coordination.

Author Contributions: All authors proposed the idea as well as contributed equally to writing the paper. H.Q.T. designed the proposed algorithm and performed the simulation. As the corresponding author, C.H. supervised the research and revised the paper.

Funding: This research received no external funding.

Conflicts of Interest: The authors declare no conflict of interest.

References

1. Kaplan, E.D.; Hegarty, C.J. *Understanding GPS: Principles and Applications*; Artech House Publishers: Norwood, MA, USA, 2006; ISBN 1-58053-894-0.
2. Zhuang, Y.; Hua, L.; Qi, L.; Yang, J.; Cao, P.; Cao, Y.; Wu, Y.; Thompson, J.; Haas, H. A Survey of Positioning Systems Using Visible LED Lights. *IEEE Commun. Surv. Tutor.* **2018**, *20*, 1963–1988. [CrossRef]
3. Chowdhury, M.Z.; Hossan, M.T.; Islam, A.; Jang, Y.M. A comparative survey of optical wireless technologies: Architectures and applications. *IEEE Access* **2018**, *6*, 9819–9840. [CrossRef]
4. Modern Indoor Living Can Be Bad for Your Health: New YouGov Survey for VELUX Sheds Light on Risks of the "Indoor Generation". Available online: https://www.prnewswire.com/news-releases/modern-indoor-living-can-be-bad-for-your-health-new-yougov-survey-for-velux-sheds-light-on-risks-of-the-indoor-generation-300648499.html (accessed on 27 November 2018).
5. Ghassemlooy, Z.; Popoola, W.; Rajbhandari, S. *Optical Wireless Communications, System and Channel Modeling with MATLAB*; CRC Press: Boca Raton, FL, USA, 2012; ISBN 9781439851883.
6. Liu, H.; Darabi, H.; Banerjee, P.; Liu, J. Survey of Wireless Indoor Positioning Techniques and Systems. *IEEE Trans. Syst. Man. Cybern. Part C (Appl. Rev.)* **2007**, *37*, 1067–1080. [CrossRef]
7. Eroglu, Y.S.; Guvenc, I.; Pala, N.; Yuksel, M. AOA-based localization and tracking in multi-element VLC systems. In Proceedings of the 2015 IEEE 16th Annual Wireless and Microwave Technology Conference (WAMICON), Cocoa Beach, FL, USA, 13–15 April 2015. [CrossRef]

8. Du, P.; Zhang, S.; Chen, C.; Alphones, A.; Zhong, W.D. Demonstration of a Low-Complexity Indoor Visible Light Positioning System Using an Enhanced TDOA Scheme. *IEEE Photonics J.* **2018**, *10*, 1–10. [CrossRef]
9. Jung, S.; Hann, S.; Park, C. TDOA-based optical wireless indoor localization using LED ceiling lamps. *IEEE Trans. Consum. Electron.* **2011**, *57*, 1592–1597. [CrossRef]
10. Zhang, S.; Zhong, W.D.; Du, P.; Chen, C. Experimental Demonstration of Indoor Sub-Decimetre Accuracy VLP System Using Differential PDOA. *IEEE Photonics Technol. Lett.* **2018**, *30*, 1703–1706. [CrossRef]
11. Naz, A.; Asif, H.M.; Umer, T.; Kim, B.S. PDOA Based Indoor Positioning Using Visible Light Communication. *IEEE Access* **2018**, *6*, 7557–7564. [CrossRef]
12. Van, M.T.; Van Tuan, N.; Son, T.T.; Le-Minh, H.; Burton, A. Weighted k-nearest neighbour model for indoor VLC positioning. *IET Commun.* **2017**, *11*, 864–871. [CrossRef]
13. Guo, X.; Shao, S.; Ansari, N.; Khreishah, A. Indoor Localization Using Visible Light via Fusion of Multiple Classifiers. *IEEE Photonics J.* **2017**, *9*, 1–16. [CrossRef]
14. Gu, W.; Kashani, M.A.; Kavehrad, M. Multipath reflections analysis on indoor visible light positioning system. *Comput. Sci.* **2015**, *57*, 13–24. [CrossRef]
15. Mohammed, N.; Elkarim, M. Exploring the effect of diffuse reflection on indoor localization systems based on RSSI-VLC. *Opt. Express* **2015**, *23*, 20297–20313. Available online: https://www.osapublishing.org/oe/abstract.cfm?uri=oe-23-16-20297 (accessed on 27 November 2018). [CrossRef] [PubMed]
16. Gu, W.; Aminikashani, M.; Deng, P.; Kavehrad, M. Impact of Multipath Reflections on the Performance of Indoor Visible Light Positioning Systems. *J. Lightw. Technol.* **2016**, *34*, 2578–2587. [CrossRef]
17. Fu, M.; Zhu, W.; Le, Z.; Manko, D. Improved visible light communication positioning algorithm based on image sensor tilting at room corners. *IET Commun.* **2018**, *12*, 1201–1206. [CrossRef]
18. Kahn, J.M.; Barry, J.R. Wireless infrared communications. *Proc. IEEE* **1997**, *85*, 265–298. [CrossRef]
19. Gfeller, F.R.; Bapst, U. Wireless in-house data communication via diffuse infrared radiation. *Proc. IEEE* **1979**, *67*, 1474–1486. [CrossRef]
20. Fan, K.; Komine, T.; Tanaka, Y.; Nakagawa, M. The effect of reflection on indoor visible-light communication system utilizing white LEDs. In Proceedings of the 5th International Symposium on Wireless Personal Multimedia Communications, Honolulu, HI, USA, 27–30 October 2002; Volume 2, pp. 611–615. [CrossRef]
21. Improving the random Forest in Python Part 1. Available online: https://towardsdatascience.com/improving-random-forest-in-python-part-1-893916666cd (accessed on 27 November 2018).
22. Pedregosa, F.; Varoquaux, G.; Gramfort, A.; Michel, V.; Thirion, B.; Grisel, O.; Blondel, M.; Prettenhofer, P.; Weiss, R.; Vanderplas, J.; et al. Scikit-learn: Machine Learning in Python. *J. Mach. Learn. Res.* **2011**, *12*, 2825–2830.

electronics

MDPI

Article

An Analysis of the Impact of LED Tilt on Visible Light Positioning Accuracy

David Plets *, Sander Bastiaens, Luc Martens and Wout Joseph

Dept. of Information Technology, Ghent University/imec, Technologiepark 126, B-9052 Ghent, Belgium;
sander.bastiaens@ugent.be (S.B.); luc1.martens@ugent.be (L.M.); wout.joseph@ugent.be (W.J.)
* Correspondence: david.plets@ugent.be; Tel.: +32-9-33-14918

Received: 28 February 2019; Accepted: 27 March 2019; Published: 1 April 2019

Abstract: Whereas the impact of photodiode noise and reflections is heavily studied in Visible Light Positioning (VLP), an often underestimated deterioration of VLP accuracy is caused by tilt of the Light Emitting Diodes (LEDs). Small LED tilts may be hard to avoid and can have a significant impact on the claimed centimeter-accuracy of VLP systems. This paper presents a Monte-Carlo-based simulation study of the impact of LED tilt on the accuracy of Received Signal Strength (RSS)-based VLP for different localization approaches. Results show that trilateration performs worse than (normalized) Least Squares algorithms, but mainly outside the LED square. Moreover, depending on inter-LED distance and LED height, median tilt-induced errors are in the range between 1 and 6 cm for small LED tilts, with errors scaling linearly with the LED tilt severity. Two methods are proposed to estimate and correct for LED tilts and their performance is compared.

Keywords: VLP; LED tilt; Visible Light Positioning; positioning; localization algorithm

1. Introduction

1.1. Introduction on Visible Light Positioning

The introduction of Light Emitting Diodes (LEDs) for traditional lighting applications has also led to new research lines in other application domains, as LEDs can be modulated to transmit information over the visible light propagation channel. Besides large interest in high-speed Visible Light Communication (VLC), another promising application is Visible Light Positioning (VLP) [1,2], where e.g., the location of a photo diode (PD) is estimated. When using Received Signal Strength Indicator (RSSI)-based positioning [3], VLP has an advantage over well-known Radio-Frequency (RF) solutions, thanks to the absence of small-scale fading effects. Research has shed light on the impact of noise on positioning performance [4], the impact of reflections [5,6], or the impact of LED power uncertainty [7]. Also the impact of receiver tilt has been characterized [8–11], and ways to compensate for this, but it remains unclear to what extent random small tilts of the LED at the ceiling impact positioning performance.

1.2. Expected Issues Considering LED Tilt

Besides tilt induced via the LED die placement and packaging, another important source of tilt is the operator suspending the LED. As suspending a LED is typically done while looking up and standing on a ladder or a hydraulic platform lift, it becomes harder for the operator to maintain a good feeling of horizontality, compared to when standing on the ground. Intuitively, one would expect that the impact of LED tilt on the positioning error strongly relates to the lateral deviation in the receiver plane, defined as the distance between the intersect of the untilted LED normal and the receiver plane (i.e., right below the LED), and the intersect of the tilted LED normal and the receiver plane (i.e., where the LED beam is directed to). This lateral deviation is equal to the tangent of the tilt angle (in radians) multiplied by the LED-PD height difference, or, for small tilt angles, the product of the LED-PD height difference and the tilt angle (in radians) itself. Figure 1a shows the value of this deviation in the receiver plane due to a tilted LED, as a function of the LED height (i.e., the height difference between the LED and the PD). For real-life industrial VLP applications, LED heights of 7 m or more are not uncommon and lead to deviations between 12 cm (for a tilt of 1°) and 61 cm (for a tilt of 5°). This paper will quantitatively and qualitatively investigate the validity of this intuitive assumption. Furthermore, it needs investigation to what extent induced errors of multiple LEDs, each with a certain (unknown) tilt, accumulate or compensate each other.

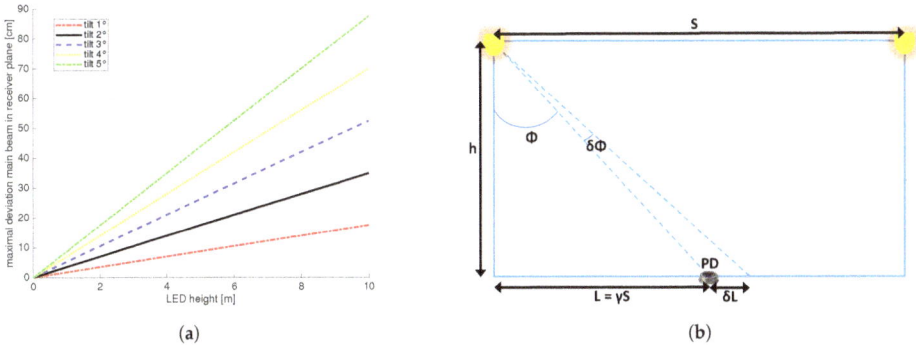

(a)   (b)

Figure 1. Lateral displacements in receiver plane in case of LED tilt. (**a**) Maximal lateral displacement of LED normal in the receiver plane due to LED tilt, as a function of LED-PD height difference; (**b**) Additional displacement ∂L in receiver plane due to LED tilt $\partial \phi$ as a function of angle of irradiance ϕ (S = square side length, h = LED-PD height difference).

1.3. Paper Content and Structure

This paper will characterize LED-tilt impact within a typical square VLP configuration by means of a Monte-Carlo simulation, in which it is assumed that each of the four deployed LEDs has a certain unknown (small) horizontal tilt, and a random azimuthal rotation. This way, the cumulative distribution function (cdf) of the positioning error will be constructed for different locations within the test site, and for different localization approaches. Also, the impact of LED tilt will be related with the LED height and inter-LED distance. Finally, a brief exploration is presented towards estimating LED tilt by investigating two tilt estimation methods, which can then be used to adjust model-based RSS-VLP fingerprinting maps, and eventually, reduce positioning errors. The remainder of this paper is structured as follows. Section 2 will present the visible light channel model that will be used in the positioning algorithm, the simulation configuration, the positioning algorithms, and the proposed LED tilt estimation methods. In Section 3, the results will be presented, after which the main findings of this work will be discussed in Section 4, together with related future research work.

2. Materials and Methods

2.1. Channel Model

In this work, only the Line-of-Sight (LoS) path between the LED transmitters and the PD receivers are accounted for. We will not consider the impact of reflections as a source of "noise", in order to be able to unambiguously assess the effect of LED tilt only. For the same reason, no shot noise or thermal noise will be considered in this study. It should be noted that even when reflections are accounted for via a model-based fingerprinting approach [6], LED tilt will have an impact since the reflected contributions could either increase or decrease, depending on the LED location and tilt, the PD location, and the location of the reflective surface. This is considered to be future work. Figure 2 and Table 1 define the model parameters of the visible light channel. The power P_R received at the PD is calculated according to the channel model used in [12]:

$$P_R = P_E \cdot h_{LoS}, \tag{1}$$

with P_E the emitted optical power by the LED. h_{LoS} is the channel gain along the direct link and can be described as follows, when assuming a Lambertian radiator:

$$h_{LoS} = \frac{m+1}{2\pi d^2} cos^m(\phi) \cdot A_R \cdot cos(\psi) \cdot T_R(\psi) \cdot G_R(\psi), \tag{2}$$

with m the order of the Lambertian emitter, and ϕ the angle of irradiance (i.e., the angle between the LED normal and the vector $\vec{v}_{LED2PD}$ from the LED to the PD, with ϕ equal to either φ_{tilt} or φ in Figure 2, depending on the LED being tilted or not). $T_R(\psi)$ and $G_R(\psi)$ are the optical filter's gain and the optical concentrator's gain at the receiver, respectively, with ψ the angle of incidence (i.e., the angle between the PD normal $\vec{n}_{PD}$ and the vector from the PD to the LED). The LEDs will be assumed to be within the field-of-view (FOV) of the PD and $T_R(\psi)$ and $G_R(\psi)$ are assumed equal to 1. d is the distance between the LED and the PD, and A_R the actual PD area, here assumed to be 1 cm^2. The PD will be assumed to be horizontally oriented, so that $cos(\psi)$ reduces to h/d, with h the height difference between the LED and the PD.

When the LED is not tilted (i.e., horizontally oriented), the angle of irradiance $\phi = \varphi = \psi$ (see Figure 2) with $cos(\varphi)$ also equal to h/d. However, in case of LED tilt, the angle of irradiance ϕ is equal to φ_{tilt} (see Figure 2), with $cos(\varphi_{tilt})$ as follows:

$$cos(\varphi_{tilt}) = \frac{\vec{v}_{LED2PD} \cdot \vec{n}_{LED}^{tilt}}{|\vec{v}_{LED2PD}| \cdot |\vec{n}_{LED}^{tilt}|}, \tag{3}$$

with $\vec{n}_{LED}^{tilt}$ the (tilted) LED normal (compared to $\vec{n}_{LED}^{no\ tilt}$, being the untilted LED normal). The LED is assumed to be tilted over an angle θ (determining the severity of the tilt), and rotated over an angle α (determining where the LED is tilted to), as shown in Figure 2. Since $|\vec{n}_{LED}^{tilt}| = 1$ and $|\vec{v}_{LED2PD}| = d$, $cos(\varphi_{tilt})$ is obtained as follows:

$$cos(\varphi_{tilt}) = \frac{(x_{PD} - x_{LED})sin(\theta)cos(\alpha) + (y_{PD} - y_{LED})sin(\theta)sin(\alpha) + h\,cos(\theta)}{d}, \tag{4}$$

with $(x_{PD}, y_{PD}, 0)$ and (x_{LED}, y_{LED}, h) the coordinates of the PD and the LED respectively. Please note that the assumption of the PD being located in the xy-plane, does not retract from the generality of the work, as only the LED-PD height difference h matters.

As such, the power P_R^{tilted} received from a tilted LED on a horizontally oriented PD is calculated as follows:

$$P_R^{tilted} = P_E \frac{h(m+1)A_R}{2\pi d^4} ((x_{PD} - x_{LED})sin(\theta)cos(\alpha) + (y_{PD} - y_{LED})sin(\theta)sin(\alpha) + hcos(\theta))^m \tag{5}$$

Table 1. Summary of the parameters defined in Figure 2.

Parameter	Explanation	Parameter	Explanation
θ	horizontal LED tilt	h	LED-PD height difference
α	azimuthal rotation of tilted LED normal	d	LED-PD distance
ϕ	general notation for angle of irradiance ($=\varphi$ or φ_{tilt})	$\vec{v}_{LED2PD}$	vector from LED to PD
φ	angle of irradiance for untilted LED	$\vec{n}_{LED}^{no\ tilt}$	untilted LED normal
φ_{tilt}	angle of irradiance for tilted LED	$\vec{n}_{LED}^{tilt}$	tilted LED normal
ψ	angle of incidence	$\vec{n}_{PD}$	PD normal

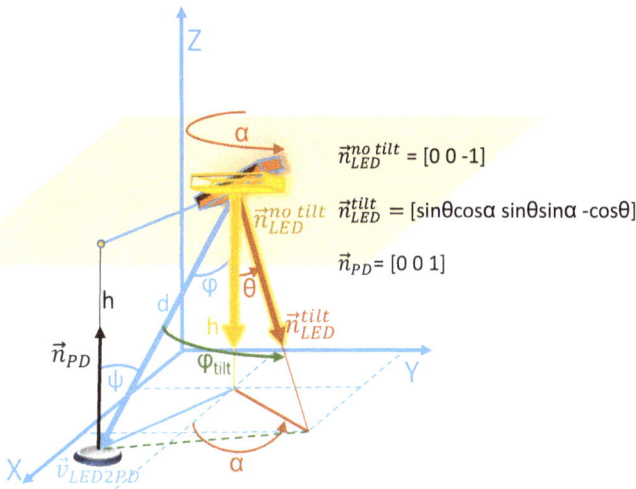

$$\vec{n}_{LED}^{no\ tilt} = [0\ 0\ {-}1]$$

$$\vec{n}_{LED}^{tilt} = [\sin\theta\cos\alpha\ \sin\theta\sin\alpha\ {-}\cos\theta]$$

$$\vec{n}_{PD} = [0\ 0\ 1]$$

Figure 2. Overview of visible light channel.

2.2. Simulation Configuration

Simulations will be executed for the room depicted in Figure 3. The dimensions of the room are 7 m × 7 m, with a ceiling at a height difference h above the PD, with $h = 2.5$ m (office environment) or $h = 6$ m (industrial environment). The Lambertian order m of the LEDs is equal to 1, for all four LEDs. In this scenario, we assume that the receiver height is fixed and known (e.g., a PD attached to the top of a cart), so the evaluation of the receiver location is reduced to a plane. A receiver grid of 5 mm will be considered here, meaning that the PD center can be located at $N_L = 1401^2$ candidate locations. Furthermore, we assume that the receiver hardware is able to demultiplex the contributions of the different LED sources [13]. Four LEDs are attached to the ceiling, at the locations indicated in Figure 3. We here assume the frequency division multiple access (FDMA) scheme proposed in [13] and further investigated in [14], which combines the transmission of square waves with 'power spectrum identification' based on the Fast Fourier Transform of the incident signal. The authors make use of the even harmonics of square waves being zero, to separate the different contributions of each LED transmitter at the receiver side. The modulation frequencies are typically in the range of 1–100 kHz, which is sufficiently below the LED and LED driver bandwidth (>1 MHz) to not have a detrimental impact on the positioning system. Although the exact optical power of the LEDs does not impact the findings, as no receiver noise is added, we here assume an optical power of 10 W, in line with common assumptions in literature [15,16].

The parameter θ is assumed to be the absolute value of a normally distributed variable θ_N, and α to be uniformly distributed: $\theta = |\theta_N|$, with $\theta_N \sim \mathcal{N}(0, \sigma_{tilt}^2)$ and $\alpha = U[0°, 360°]$. Two values of σ_{tilt} will be mainly considered in this work: 1° and 2°, meaning that 95% of the LEDs will be mounted with a tilt below 2° and 4° respectively.

The positioning error due to tilt will be evaluated at 100 locations in a quarter of the receiver plane, i.e., at the blue locations indicated in Figure 3. Thanks to the symmetry of the setup, the resulting error distributions can be extrapolated to the three other zones, since the distribution of the positioning errors will repeat themselves at the corresponding locations of the different parts of the 7 m × 7 m area. It should be noted though that each single simulation will result in an asymmetric setup, since for each simulation, the tilt of each LED will be randomly and independently chosen. However, the resulting cdf at corresponding locations will be the same when a sufficient number of simulations is considered, due to the fact that the statistical distribution of the LED tilt is assumed to be the same for each of the LEDs. In this work, a position estimation for each of the 100 positions will be executed for 10,000 random sets of LED tilts in a Monte-Carlo simulation. Each position estimation will be done according to the algorithms presented in Section 2.3.

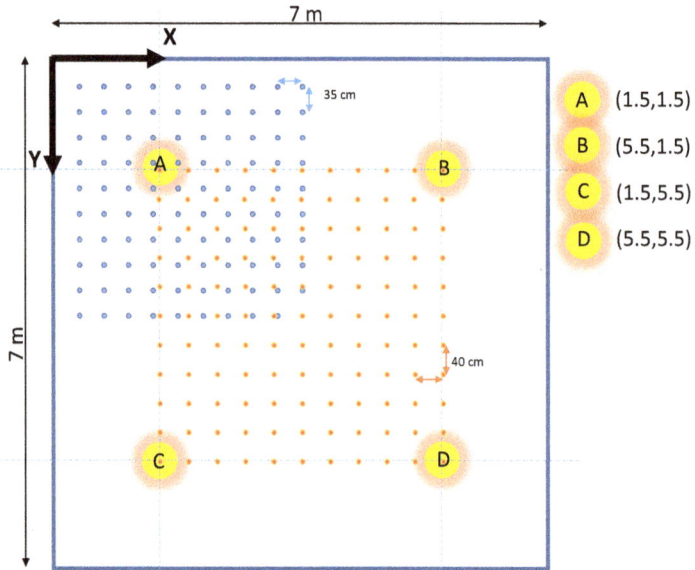

Figure 3. Simulation configuration (A,B,C,D indicate LED locations, blue dots indicate scatter plot ground truth locations, see Section 3.1, orange dots indicate the evaluation points used in Section 3.2).

2.3. Positioning Algorithms

Three different positioning algorithms will be compared: a traditional trilateration method, and two model-based fingerprinting methods using a Least-Squares Estimation (LSE) and a normalized LSE (nLSE) respectively.

2.3.1. Trilateration

Given a measured power P_{Ri}^{meas} from LED_i ($i = 1..N$) and assuming that $cos(\phi) = cos(\psi) = \frac{h}{d_i}$ for horizontally oriented LEDs and PD (see Figure 2), Equation (2) can be rewritten to allow the calculation of the estimated distance $\hat{d}_i$ between LED_i and PD:

$$\hat{d}_i = \sqrt[m+3]{(m+1) \cdot \frac{P_E}{2\pi P_{Ri}^{meas}} \cdot h^{m+1} \cdot A_R,} \tag{6}$$

The real squared horizontal distance d_i^2 between LED_i and the PD is given by:

$$d_i^2 = (x - x_i)^2 + (y - y_i)^2 = x^2 - 2xx_i + x_i^2 + y^2 - 2yy_i + y_i^2, \tag{7}$$

with (x, y) the PD location, and (x_i, y_i) the coordinates of LED_i. After eliminating the quadratic terms in x^2 and y^2 by subtracting d_N^2 from d_i^2, N-1 equations are obtained ($i = 1, \ldots, N - 1$):

$$d_i^2 - d_N^2 = -2x(x_i - x_N) + x_i^2 - x_N^2 - 2y(y_i - y_N) + y_i^2 - y_N^2 \tag{8}$$

These linear equations in x and y can be written as $b = M \begin{bmatrix} x \\ y \end{bmatrix}$, where

$$
b = \begin{bmatrix} d_1^2 - x_1^2 - y_1^2 - d_N^2 + x_N^2 + y_N^2 \\ d_2^2 - x_2^2 - y_2^2 - d_N^2 + x_N^2 + y_N^2 \\ \vdots \\ d_{N-1}^2 - x_{N-1}^2 - y_{N-1}^2 - d_N^2 + x_N^2 + y_N^2 \end{bmatrix} \tag{9}
$$

$$
M = \begin{bmatrix} x_1 - x_N & y_1 - y_N \\ x_2 - x_N & y_2 - y_N \\ \vdots & \vdots \\ x_{N-1} - x_N & y_{N-1} - y_N \end{bmatrix} \tag{10}
$$

$\begin{bmatrix} x \\ y \end{bmatrix}$ can then be estimated as $\begin{bmatrix} \hat{x} \\ \hat{y} \end{bmatrix}$ using the Moore-Penrose pseudo-inverse of M:

$$
\begin{bmatrix} \hat{x} \\ \hat{y} \end{bmatrix} = (M^T M)^{-1} M^T b \tag{11}
$$

In case the LEDs are tilted, the assumption of $cos(\phi) = cos(\psi) = \frac{h}{d_i}$ will no longer hold, and errors will be introduced.

2.3.2. Least-Squares Estimation

The following two methods are based on the comparison of the set of so-called measured received powers P_{Ri}^{meas} from each (tilted) LED_i ($i = 1, \ldots, N$) at the unknown PD location (x_{PD}, y_{PD}), with the set of fingerprinted PD powers $P_{Ri}^{L,model}$ from LED_i at all (1401^2) locations L in the grid. For the construction of the fingerprinting database of the $P_{Ri}^{L,model}$ values, each LED is assumed to be untilted, as it is its most probable position. The set of so-called measurements $(P_{R1}^{meas}, P_{R2}^{meas}, \ldots, P_{RN}^{meas})$ represent the observed values in the realistic setup investigated here. They are obtained from $(P_{E1}, P_{E2}, \ldots, P_{EN})$, where P_{Ri}^{meas} values are obtained from Equation (5) with θ and α values as samples from their respective statistical distributions. The larger the tilt of the LEDs (larger σ_{tilt}^2 values), the larger the positioning errors will be. The algorithm estimates the unknown location with coordinates (x_{PD}, y_{PD}) to be at the location L where the Least-Squares cost function C_{LSE}^L has a minimum [6]:

$$
C_{LSE}^L = \sum_{i}^{N} (P_{Ri}^{meas} - P_{Ri}^{L,model})^2. \tag{12}
$$

2.3.3. Normalized Least-Squares Estimation

The normalized Least-Squares Estimation algorithm was shown to perform better than the LSE algorithm when there is uncertainty on the respective LED powers [17]. The unknown location (x_{PD}, y_{PD}) is estimated at the the location L where the cost function C_{nLSE}^L has a minimum:

$$
C_{nLSE}^L = \sum_{i}^{N} \left(\frac{P_{Ri}^{meas} - P_{Ri}^{L,model}}{P_{Ri}^{L,model}} \right)^2. \tag{13}
$$

For the LSE and nLSE methods, each position estimation thus consists of a comparison of the set of measurements $(P_{R1}^{meas}, P_{R2}^{meas}, \ldots, P_{RN}^{meas})$ against all $(P_{R1}^{L,model}, P_{R2}^{L,model}, \ldots, P_{RN}^{L,model})$ sets that are stored in the database. In total, N_L sets of N values of $P_{Ri}^{L,model}$ are precalculated and stored in a fingerprinting database, i.e., the received power at N_L locations from each of the N LEDs, according to the LoS channel model from Section 2.1. For the configuration under test, N = 4 and $N_L = 1401^2 = 1,962,801$, meaning that 7,851,204 values are stored.

2.4. LED Tilt Estimation Methods

Based on the configuration and the algorithms described in the previous sections, the impact of LED tilt will be assessed. In this section, two methods for estimating the LED tilt are presented, i.e., the angles θ and α from Figure 2. Once this tilt is known, the fingerprinting map with powers ($P_{Ri}^{L,model}$ values of Sections 2.3.2 and 2.3.3) can be adjusted, in order to reduce aforementioned LED tilt impact.

2.4.1. Exhaustive Search

The first method encompasses an exhaustive search over all possible (x,y) locations in the receiver plane. As Equation (2) can be rewritten as

$$P_R(x,y) = P_E \cdot \frac{m+1}{2\pi d^2(x,y)} \cos^m(\phi)(x,y) \cdot A_R \cdot \cos(\psi)(x,y) \cdot T_R(\psi)(x,y) \cdot G_R(\psi)(x,y), \quad (14)$$

the location (x_L,y_L) in the receiver plane which is pointed to by the LED normal, can be determined as the (x,y) location for which the angle of irradiance ϕ is minimal or $\cos^m(\phi)(x,y)$ is maximal. For a horizontal PD, and when removing all (x,y)-independent factors, (x_L,y_L) is thus found at the location (x,y) where

$$\frac{P_R(x,y) \cdot d^3(x,y)}{T_R(\psi)(x,y) \cdot G_R(\psi)(x,y)} \quad (15)$$

is maximal. Especially for small LED tilt values, it is fair to assume that the optical filter's gain and the optical concentrator's gain at the receiver are also independent of the receiver location, so that the LED tilt is determined by the location (x_L,y_L) where the product of the measured received power $P_R(x,y)$ and the cube of the LED-PD distance becomes maximal, i.e.,

$$(x_L,y_L) = \max_{(x,y)} (P_R(x,y) \cdot d^3(x,y)).$$

From the location (x_L,y_L), the tilt values (α,θ) are easily derived using basic trigonometry as ($\arctan \frac{y_L}{x_L}$, $\arctan \frac{\sqrt{x_L^2+y_L^2}}{h}$). Although this method in principle delivers the exact LED tilt, it is very cumbersome, as it requires the execution of a large set of measurements, whereby the exact ground truth of the measurement location has to be known. The measurement campaign should be executed with an automated positioning system, e.g., with a stepping motor [18].

2.4.2. Quick Search

Unlike in the case of the first method, the second method, denoted here as 'quick search', uses a power measurement at only a limited set of N_M locations under the considered LED. Based on these power measurements P_i^{meas}, a minimum-search in the (θ,α) space is executed for finding the most likely values for these tilt parameters. We propose a cost function using power ratios at the different locations instead of absolute power values, since the expected powers P_i^{model} are sometimes not known without doing a calibration phase like e.g., in [19]. This is due to e.g., unknown deviations of the assumed LED power, PD responsivity,... When working with power ratios and thus excluding these unknown factors to a certain extent, the cost function is expected to be better suited to identify power differences that are solely due to the LED tilt. As such, the proposed cost function is the following:

$$C_{tilt}(\alpha,\theta) = \sum_{i=1}^{N_M} \sum_{j=i+1}^{N_M} \left(\frac{P_i^{meas}}{P_j^{meas}} - \frac{P_i^{model}(\alpha,\theta)}{P_j^{model}(\alpha,\theta)} \right)^2, \quad (16)$$

with N_M the number of measurement locations in the receiver plane, and P_i^{meas} the measured power at location i underneath the considered LED. The power values P_i, i=1..N_M, are first sorted in descending order. $P_i^{model}(\alpha, \theta)$ is the modeled received power at location i, according to Equation (5) for a LED tilt determined by (α, θ). The cost function iterates over α between 0° and 360°, with a 1° resolution. The parameter θ is iterated between 0° and $\theta_{max} = 5$, with 0.1° resolution. C_{tilt} will reach its minimal value for (α, θ) values equal to the actual LED tilt values, although the outcome will be corrupted by external factors such as noise, unaccounted PD tilt, deviations from the tabulated Lambertian order.

2.4.3. LED Tilt Estimation Scenario

The performance of the described methods will be evaluated for a 10 W LED at a height of 2.5 m above the receiver plane. The LED is assumed to have a horizontal tilt of $\theta = 1.6°$ and an azimuthal rotation α of 230°. For the 'quick search' method, four measurement locations ($N_M = 4$) in the receiver plane will be considered, with (quite randomly chosen) coordinates (1,0), $(-0.5, \pm\frac{\sqrt{3}}{2})$, and (1, −1). Finally, a PD area of $A_R = 1$ cm^2 is assumed.

3. Results

3.1. Assessment of LED Tilt Impact on Positioning Accuracy for Typical Configurations Using Different Metrics

Figure 4 shows the cdf of the positioning errors of the 10^4 simulations at each of the 100 evaluation locations for (a) a LED-PD height difference of 2.5 m and $\sigma_{tilt} = 1°$ (denoted as normal office placement); (b) a LED-PD height difference of 2.5 m and $\sigma_{tilt} = 2°$ (denoted as sloppy office placement); (c) a LED-PD height difference of 6 m and $\sigma_{tilt} = 1°$ (denoted as normal industrial placement); and (d) a LED-PD height difference of 6 m and $\sigma_{tilt} = 2°$ (denoted as sloppy industrial placement). In each plot, the three localisation metrics from Section 2.3 are compared. Table 2 lists the median (p_{50}) and 95%-percentile (p_{95}) values of the errors for the different configurations. It is observed that for a normal office placement (see Figure 4a), median errors are around 2–3 cm and 95-percentile errors between 6 and 10 cm. The LSE metric slightly outperforms the nLSE metric, whereas traditional trilateration produces errors that are almost 40% higher on average than for LSE. For a sloppy industrial placement (see Figure 4b), median (p_{50}) and maximal (p_{95}) errors more or less double, compared to the normal placement, for all metrics.

When increasing the height to 6 m (industrial), errors increase, but not significantly. For a normal industrial placement (see Figure 4c), although median errors increase between 8 (trilateration) and 24% (LSE) compared to the normal office placement, maximal errors even see a slight decrease for the LSE and trilateration approaches. The next section will elaborate on this impact of LED height more thoroughly. Finally, for a sloppy industrial placement (see Figure 4d), the errors again more or less double compared to the normal industrial placement, suggesting that doubling σ_{tilt} for each of the LEDs also doubles the resulting positioning error.

To give more insight on the spatial distribution of the postion estimates for the different localization approaches, Figure 5a–c show a scatter plot of the estimated positions for a LED-PD height difference of 2.5 m, and a σ_{tilt} value of 2°, for LSE, nLSE, and trilateration, respectively. For each of the 100 evaluation points, 250 estimations are displayed. These 100 evaluation points correspond to the blue dots in Figure 3. The square that is formed by the LEDs is partly shown with orange lines. These scatter plots correspond to the values in Table 2 for a 'sloppy office' deployment. The dominant LED at (1.5, 1.5) causes a more circular pattern due its LED tilt for the LSE metric than for nLSE, showing that nLSE indeed reduces the impact of the dominant LED. Trilateration performs worse than the other two metrics, but especially outside the LED square (i.e., for x or y coordinates <1.5 m). Inside the LED square, the performance of the algorithms is comparable.

Table 2. Median and maximal positioning errors for the three localization approaches, for four typical LED deployments.

Positioning Error (cm)	LSE		nLSE		Trilateration	
	p_{50}	p_{95}	p_{50}	p_{95}	p_{50}	p_{95}
normal office ($h = 2.5$, $\sigma_{tilt} = 1°$)	1.80	6.52	2.06	6.52	2.47	9.07
sloppy office ($h = 2.5$, $\sigma_{tilt} = 2°$)	3.64	13.09	4.03	13.15	4.90	18.35
normal industrial ($h = 6$, $\sigma_{tilt} = 1°$)	2.24	6.40	2.50	7.02	2.66	8.08
sloppy industrial ($h = 6$, $\sigma_{tilt} = 2°$)	4.53	12.65	4.92	13.87	5.25	15.87

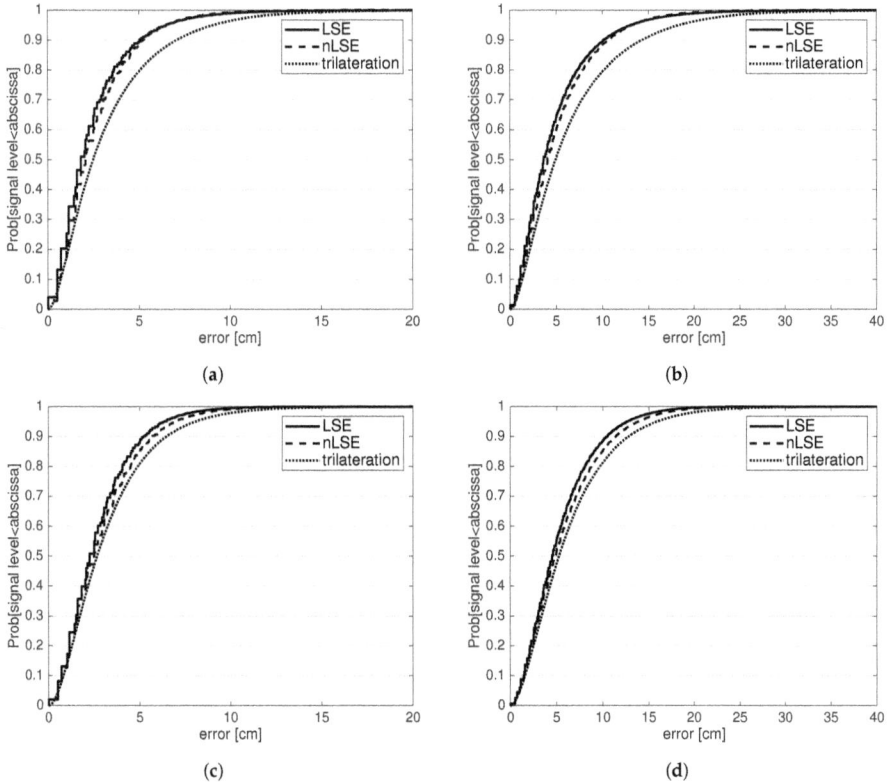

(a)

(b)

(c)

(d)

Figure 4. Cdf of the positioning errors at the evaluation locations (blue dots in Figure 3) for three positioning metrics. (**a**) Normal office placement (LED-PD height difference = 2.5 m, $\sigma_{tilt} = 1°$); (**b**) Sloppy office placement (LED-PD height difference = 2.5 m, $\sigma_{tilt} = 2°$); (**c**) Normal industrial placement (LED-PD height difference = 6 m, $\sigma_{tilt} = 1°$); (**d**) Sloppy industrial placement (LED-PD height difference = 6 m, $\sigma_{tilt} = 2°$).

(a) LSE

(b) normalized LSE

(c) Trilateration

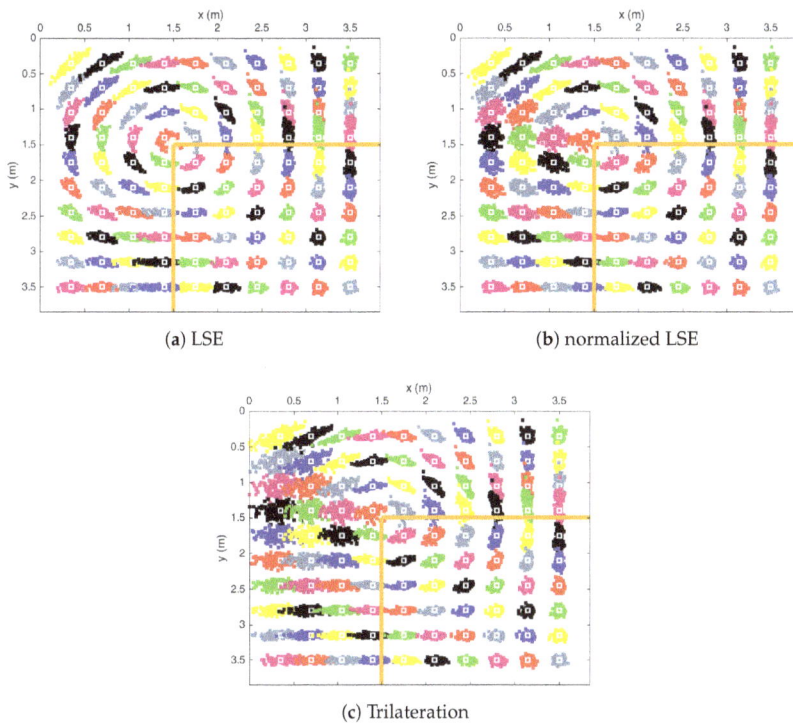

Figure 5. Scatter plot of 250 location estimations for each of 100 locations (indicated by white-edged dots) in top left part of the 7 m by 7 m square of Figure 3, for a LED height of 2.5 m and a $\sigma_{tilt} = 2°$. The LED square is indicated with an orange line.

3.2. LED Tilt Impact for Different Inter-LED Distances, LED Height and Tilt

In this section, we further characterize the errors due to random LED tilt, using the common trilateration method described in Section 2.3.1. Different configurations with four LEDs in a square configuration are investigated. We vary the side of the square (i.e., the inter-LED distance) between 2 and 8 m in steps of 1 m, the LED height between 2 and 8 m in steps of 1 m, and σ_{tilt} between 1 and 3° in steps of 1°. For each of these $7 \times 7 \times 3 = 147$ combinations, each of the 4 LEDs are randomly tilted and the localisation accuracy is evaluated on a uniform 11×11 grid in between the LED locations, meaning that the accuracy is evaluated only inside the LED square, for which was observed that all three localization approaches yielded comparable estimation clouds (see Figure 5). These 121 evaluation points correspond to the orange dots in Figure 3. Please note that Figure 3 represents a configuration with square side length $S = 4$. For each of the 147 combinations, 5000 LED tilt settings are generated, with the LEDs tilted according to the statistical tilt distributions presented in Section 2.2. As (thermal or shot) noise is not considered here, the value of the transmit power P_E or the PD area A_R has no influence on the positioning accuracy, so conclusions are also valid for other LED optical powers or PD areas.

Figure 6a shows the p_{50} error due to LED tilt for 7 different square side lengths (2 to 8 m), as a function of the LED height, for a σ_{tilt} value of 1°. The figure shows that irrespective of the LED height, the impact of tilt is the lowest for smaller LED squares. e.g., for a LED height of 2 m, the median induced positioning error increases from 1 cm for a 2×2 m LED square to 6.5 cm for an 8×8 m LED square. As the LED height increases, the influence of the square side length reduces, with median errors between 2.4 cm (2×2 square) and 3.8 cm (8×8 m square). Furthermore, it is interesting to note that the impact of LED tilt on positioning accuracy in the receiver plane does not necessarily increase with LED height (i.e., higher LED-PD height differences), while intuitively (and from Figure 1a), one might expect a monotonic increase of the induced error with LED height. However, for a given LED square size S, there appears to be an optimal LED height with respect to the impact of LED tilt. e.g., for an 8×8 m LED square, this optimal LED height is 6 m, decreasing to 4 m for a 5×5 m LED square, and to smaller than 2 m for a 2×2 m LED square.

We explain this using Figure 1b, where a 1D-simplified side view of the considered configuration is shown, with h the LED-PD height difference and S the LED square side length. With γ between 0 and 1, γS indicates the range of possible lateral displacements $L = h \cdot tan\phi$ of the PD in the receiver plane, with respect to perpendicular irradiance ($\gamma = 0$ is below the LED, $\gamma = 1$ is below an adjacent LED of the LED square). ϕ is the angle of irradiance for an untilted LED. The variable of interest here is $\frac{\partial L}{\partial \phi}$, indicating the additional lateral displacement in the receiver plane, due to a LED tilt $\partial \phi$: $\frac{\partial L}{\partial \phi} = \frac{h}{cos^2\phi} = h + \frac{\gamma^2 S^2}{h}$. The result comprises two opposed phenomena. One the one hand, we observe that larger LED heights h correspond to larger additional displacements in the receiver plane. This corresponds to the phenomenon described in Section 1.2 and Figure 1, and is is reflected by the first factor h of $\frac{\partial L}{\partial \phi}$, meaning that the tilt impact is smaller for smaller LED heights. On the other hand, we see that for a given (and fixed) LED square size S, relatively more receiver locations will have a large angle of irradiance ϕ when h is smaller (see also Figure 1b), leading to a larger additional lateral displacement. This is reflected by the second factor $\frac{\gamma^2 S^2}{h}$ of $\frac{\partial L}{\partial \phi}$. The combination of these two opposed phenomena causes, depending on the size of the LED square S, the minimal tilt impact to be observed at different height differences h. When the LED-PD height difference is equal to the inter-LED distance ($h = S$), forming a cube, the error sees a perfect linear increase with h, as indicated by the red curve in Figure 6a, again corresponding to the intuitive assumption described in Section 1.2 ($\frac{\partial L}{\partial \phi}$ reduces to $h(1 + \gamma^2)$ for this simplified 1D-case).

Finally, it was again observed that p_{95} errors scale approximately to values equal to three times the p_{50} values. Similarly, the p_{50} errors scale also linearly with the σ_{tilt} value, as is shown in Figure 6b, depicting the induced median error for a 4×4 m LED square, as a function of LED height for three σ_{tilt} values.

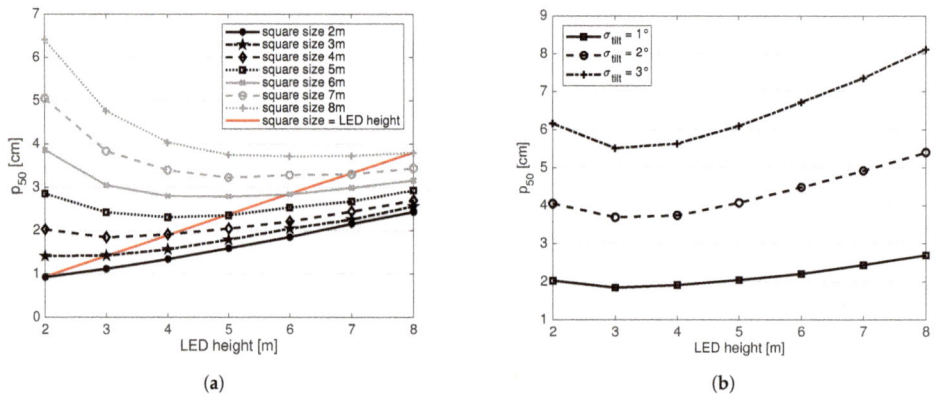

Figure 6. Induced median error due to LED tilt for different LED square sizes, heights, and tilts. (a) Induced median error for square side lengths between 2 and 8 m, as a function of LED height, for a σ_{tilt} value of 1°; (b) Induced median error for a 4×4 m LED square, as a function of LED height for three σ_{tilt} values.

3.3. Evaluation of LED Tilt Estimation Methods

In this section, the performance of the LED tilt estimation methods presented in Section 2.4 are presented. Figure 7a,b show the $P_R \cdot d^3$ values in the receiver plane without receiver noise and with added noise to the received powers ($\sigma_{noise} = 10^{-7}$ W) respectively ('exhaustive search method'). The figures indicate the LED location (green asterisk in (0,0)), the real (black asterisk) and estimated (red asterisk) intersect of the LED normal with the receiver plane, and the location receiving the maximal power P_R (blue asterisk). It should be noted that the direction of the tilted LED normal is indeed not only determined by the maximal received power P_R (blue asterisk), as this maximal-power location is determined by a tradeoff between being located along the LED normal (pulling towards the black asterisk) and having the shortest distance to the LED (pulling the location back towards right underneath the LED, green asterisk). This is also why the tilted LED normal intersect is not found at the location with a maximal P_R, but instead at the location with a maximal $P_R \cdot d^3$. Figure 7a shows that under the absence of noise, the LED tilt can be estimated exactly (red asterisk co-located with black asterisk), indicating the correctness of the proposed method. When noise is added, θ and α are estimated at 2.1° and 201° respectively (compared to the real values $\theta = 1.6°$ and $\alpha = 230°$). To reduce or avoid the possible effect of obtaining an erroneous noise-induced maximum of $P_R \cdot d^3$, an alternative approach could be to calculate the centre of the set of the X locations with the highest measured $P_R \cdot d^3$ values. The value of X should be determined based on a tradeoff between a sufficient tilt estimation precision and a sufficient spatial averaging of possible noise.

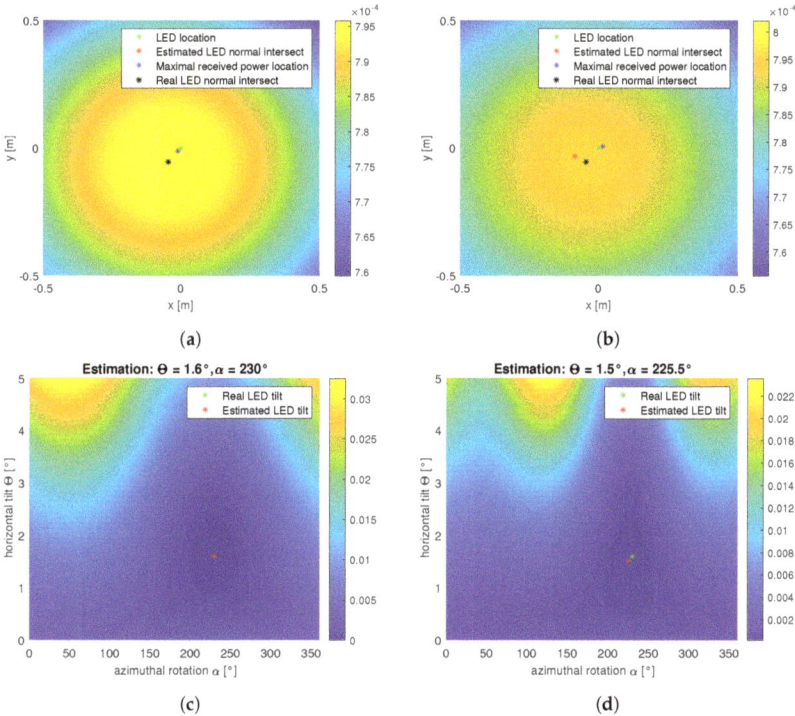

Figure 7. Illustration of LED tilt estimation methods. (**a**) Exhaustive method—no noise (scale showing $P_R \cdot d^3$, in Wm^3); (**b**) Exhaustive method—noise (scale showing $P_R \cdot d^3$, in Wm^3); (**c**) Quick search—no noise (scale showing $C_{tilt}(\alpha, \theta)$ from Equation (16), dimensionless); (**d**) Quick search—noise (scale showing $C_{tilt}(\alpha, \theta)$ from Equation (16), dimensionless).

Figure 7c,d show the cost function from Equation (16) without noise and with added noise to the considered measurement points ($\sigma_{noise} = 10^{-7}$ W). Figure 7c indicates the correct operation of the quick search method (green asterisk co-located with red asterisk). Figure 7d shows a slight deviation of the estimation due to the added noise (error in θ and α of 0.1° and 4.5° respectively). Comparison of the LED tilt estimations of the two methods under the presence of noise, shows that the quick search using only four measurement points typically performs better than the exhaustive search, despite the fact that fewer measurements were used. This is explained by the fact that the intersects of the tilted LED normals are usually very close to the untilted LED normal (a few cm, see Figure 7a,b). Due to the low gradient of the received power in the area underneath the LED (m = 1), differences in P_R (or in $P_R \cdot d^3$) are very small, making them susceptible to noise. The quick search method uses measurement points that are further from the intersect of the untilted LED normal with the receiver plane. Thanks to the larger power gradient there, this method appears to be less susceptible to noise. Or differently stated, LED tilt is better noticed at larger angles of irradiance, the phenomenon also discussed in the previous section and noticed in Figure 1b.

4. Discussion

In this paper, we investigated to what extent the uncertainty on the LED tilt impacts RSS-based VLP accuracies. For a 7 m × 7 m room with four LEDs placed in a 4 m × 4 m square, positioning errors at 100 locations in the receiver plane were compared for three localization approaches, based on a Monte Carlo simulation consisting of 10^4 simulations. It was shown the model-based fingerprinting methods (Least-Squares Estimation and normalized Least-Squares Estimation) performed slightly better than a traditional trilateration, mainly for locations outside the LED square. We observed that tilt-induced errors are in the order of centimeters, depending on the LED configuration (LED height and inter-LED distance), and the severity of the tilt. The errors scale linearly with the severity of the LED tilt.

We found that for a σ_{tilt} value of 1°, median errors for a LED height of 7 m were between 2 and 4 cm (see Figure 6a, depending on inter-LED distance), which indicates a more limited impact than intuitively assumed from Figure 1a, where a lateral deviation of 12.2 cm was found. This indicates that the lateral deviation of the LED normal in the receiver plane corresponds to a significant overestimation of tilt-induced positioning errors, suggesting that induced errors of multiple LEDs compensate each other. It has also been shown that increasing the LED height does not necessarily increase the tilt-induced errors: depending on the inter-LED distance, there is an LED height optimum where LED tilt impact is minimal.

We now compare these errors with errors induced by noise, and errors due to deviations on the tabulated optical LED power. In [6], the performance of VLP under the presence of reflections was investigated for h = 1.65 m and S = 2.5 m and a wall reflectance factor of 0.3. Median errors between 6.7 and 8.7 cm were found, depending on the metric used. Based on the curves presented in Figure 6a, a median error between 1 and 1.5 cm due to tilt can be expected for a configuration with these h and S values, when σ_{tilt} = 1°. In [17], it was investigated to what extent deviations from the tabulated transmitted optical LED power impacted positioning accuracy. Again, for h = 1.65 m and S = 2.5 m, median and maximal (95% percentile) errors of 4.21 and 10.75 cm were found for a LED tolerance (i.e., 3-sigma value) of 10%. For the same configuration, median errors ranging from about 1 mm to 10 cm were obtained for standard deviations of the observed noise power ranging between 10^{-8} and 10^{-6} W [17].

Knowing that the tilt-induced errors investigated here, add up with those introduced by thermal noise and shot noise, reflections, receiver tilt, and imperfections in the LED radiation pattern, LED tilt is one of the crucial aspects to consider and compensate for, since the other aspects are often harder to compensate for: receiver tilt might be variable while moving, noise is a random process, and reflections and LED radiations patterns are not easy to model well, or require the execution of a large measurement campaign. Therefore, estimating LED tilt and compensating for it might consistently reduce VLP errors by a few centimeters. As a first exploration, two methods to estimate the LED tilt have been presented. Although intuitively it could be assumed that an exhaustive analysis of the (maximal) powers in the area right underneath the tilted LED would be the best approach to accurately estimate LED tilts, it seems that a limited set of measurements at different, more distant points would deliver better estimates. It remains to be (experimentally) investigated how many points would ideally be required and which locations would be optimal. This is considered as future work. The presented method shows to have the potential to estimate LED tilt, which in turn allows adjusting model-based fingerprinting maps for an improved RSS-based VLP performance. Another interesting future research track is a full sensitivity analysis, including the impact of noise, LED tilt, PD tilt, and wall reflections.

Author Contributions: Conceptualization, D.P.; Data curation, D.P. and S.B.; Funding acquisition, D.P.; Investigation, D.P. and S.B.; Methodology, D.P.; Project administration, D.P.; Resources, D.P.; Software, D.P.; Supervision, L.M. and W.J.; Validation, D.P. and S.B.; Visualization, D.P.; Writing—original draft, D.P.; Writing—review & editing, D.P., L.M. and W.J.

Funding: This work was executed within LEDsTrack, a research project bringing together academic researchers and industry partners. The LEDsTrack project was co-financed by imec (iMinds) and received project support from Flanders Innovation & Entrepreneurship.

Conflicts of Interest: The authors declare no conflict of interest.

Abbreviations

The following abbreviations are used in this manuscript:

MDPI	Multidisciplinary Digital Publishing Institute
LED	Light Emitting Diode
VLP	Visible Light Positioning
RSS	Received Signal Strength
PD	photodiode
VLC	Visible Light Communication
RSSI	Received Signal Strength Indicator
RF	Radio-frequency
cdf	cumulative distribution function
LoS	Line-of-Sight
LSE	Least-Squares Estimator
nLSE	normalized Least-Squares Estimator

References

1. Armstrong, J.; Sekercioglu, Y.A.; Neild, A. Visible Light Positioning: A Roadmap for International Standardization. *IEEE Commun. Mag.* **2013**, *51*, 68–73. [CrossRef]
2. Jovicic, A.; Li, J.; Richardson, T. Visible light communication: opportunities, challenges and the path to market. *IEEE Commun. Mag.* **2013**, *51*, 26–32. [CrossRef]
3. Trogh, J.; Plets, D.; Martens, L.; Joseph, W. Advanced Real-Time Indoor Tracking Based on the Viterbi Algorithm and Semantic Data. *Int. J. Distrib. Sens. Netw.* **2015**, *11*. [CrossRef]
4. Mousa, F.I.K.; Almaadeed, N.; Busawon, K.; Bouridane, A.; Binns, R.; Elliott, I. Indoor visible light communication localization system utilizing received signal strength indication technique and trilateration method. *Opt. Eng.* **2018**, *57*, 016107. [CrossRef]
5. Gu, W.; Aminikashani, M.; Deng, P.; Kavehrad, M. Impact of Multipath Reflections on the Performance of Indoor Visible Light Positioning Systems. *J. Lightwave Technol.* **2016**, *34*, 2578–2587. [CrossRef]
6. Plets, D.; Eryildirim, A.; Bastiaens, S.; Stevens, N.; Martens, L.; Joseph, W. A performance comparison of different cost functions for RSS-based visible light positioning under the presence of reflections. In Proceedings of the 4th ACM Workshop on Visible Light Communication Systems at the 23rd Annual International Conference on Mobile Computing and Networking, Snowbird, UT, USA, 16 October 2017; ACM Press: New York, NY, USA, 2017; pp. 37–41.
7. Plets, D.; Bastiaens, S.; Stevens, N.; Martens, L.; Joseph, W. Monte-Carlo Simulation of the Impact of LED Power Uncertainty on Visible Light Positioning Accuracy. In Proceedings of the 11th International Symposium on Communication Systems, Networks & Digital Signal Processing, CSNDSP 2018, Budapest, Hungary, 18–20 July 2018; pp. 1–6. [CrossRef]
8. Jeong, E.; Yang, S.; Kim, H.; Han, S. Tilted receiver angle error compensated indoor positioning system based on visible light communication. *Electron. Lett.* **2013**, *49*, 890–892. [CrossRef]

9. Jeong, E.M.; Kim, D.R.; Yang, S.H.; Kim, H.S.; Son, Y.H.; Han, S.K. Estimated position error compensation in localization using visible light communication. In Proceedings of the 2013 Fifth International Conference on Ubiquitous and Future Networks (ICUFN), Da Nang, Vietnam, 2–5 July 2013; pp. 470–471. [CrossRef]

10. Yang, S.; Kim, H.; Son, Y.; Han, S. Three-Dimensional Visible Light Indoor Localization Using AOA and RSS With Multiple Optical Receivers. *J. Lightwave Technol.* **2014**, *32*, 2480–2485. [CrossRef]

11. Wang, J.; Li, Q.; Zhu, J.; Wang, Y. Impact of receiver's tilted angle on channel capacity in VLCs. *Electron. Lett.* **2017**, *53*, 421–423. [CrossRef]

12. Komine, T.; Nakagawa, M. Fundamental analysis for visible-light communication system using LED lights. *IEEE Trans. Consum. Electron.* **2004**, *50*, 100–107. [CrossRef]

13. Lausnay, S.D.; Strycker, L.D.; Goemaere, J.P.; Stevens, N.; Nauwelaers, B. A Visible Light Positioning system using Frequency Division Multiple Access with square waves. In Proceedings of the 2015 9th International Conference on Signal Processing and Communication Systems (ICSPCS), Cairns, QLD, Australia, 14–16 December 2015; pp. 1–7.

14. Bastiaens, S.; Plets, D.; Martens, L.; Joseph, W. Impact of Nonideal LED Modulation on RSS-based VLP Performance. In Proceedings of the 2018 IEEE 29th Annual International Symposium on Personal, Indoor and Mobile Radio Communications (PIMRC), Bologna, Italy, 9–12 September 2018; pp. 1–5. [CrossRef]

15. Zhou, Z.; Kavehrad, M.; Deng, P. Indoor positioning algorithm using light-emitting diode visible light communications. *Opt. Eng.* **2012**, *51*, 1–7. [CrossRef]

16. Sun, X.; Duan, J.; Zou, Y.; Shi, A. Impact of multipath effects on theoretical accuracy of TOA-based indoor VLC positioning system. *Photon. Res.* **2015**, *3*, 296–299. [CrossRef]

17. Plets, D.; Bastiaens, S.; Stevens, N.; Martens, L.; Joseph, W. On the Impact of LED Power Uncertainty on the Accuracy of 2D and 3D Visible Light Positioning. *Optik* **2019**, submitted.

18. Hanssens, B.; Plets, D.; Tanghe, E.; Oestges, C.; Gaillot, D.P.; Lienard, M.; Li, T.; Steendam, H.; Martens, L.; Joseph, W. An Indoor Variance-Based Localization Technique Utilizing the UWB Estimation of Geometrical Propagation Parameters. *IEEE Trans. Antennas Propag.* **2018**, *66*, 2522–2533. [CrossRef]

19. Kim, H.; Kim, D.; Yang, S.; Son, Y.; Han, S. An Indoor Visible Light Communication Positioning System Using a RF Carrier Allocation Technique. *J. Lightwave Technol.* **2013**, *31*, 134–144. [CrossRef]

electronics

MDPI

Article

Feedforward Control Based on Error and Disturbance Observation for the CCD and Fiber-Optic Gyroscope-Based Mobile Optoelectronic Tracking System

Yong Luo [1,2,3], Wei Ren [1,2,3], Yongmei Huang [1,2], Qiunong He [1,2,3], Qiongyan Wu [1,2], Xi Zhou [1,2] and Yao Mao [1,2,*]

[1]   Institute of Optics and Electronics, Chinese Academy of Science, Chengdu 610209, China; ly250047087@126.com (Y.L.); renwei9327@163.com (W.R.); huangym@ioe.ac.cn (Y.H.); hans@myshworks.com (Q.H.); wuqiongyan@ioe.ac.cn (Q.W.); zhouxiee@mail.ustc.edu.cn (X.Z.)
[2]   Key Laboratory of Optical Engineering, Chinese Academy of Sciences, Chengdu 610209, China
[3]   University of Chinese Academy of Sciences, Beijing 100039, China
[*]   Correspondence: maoyao@ioe.ac.cn; Tel.: +86-135-4787-8788

Received: 23 August 2018; Accepted: 26 September 2018; Published: 29 September 2018

Abstract: In the mobile optoelectronic tracking system (MOTS) based on charge-coupled device (CCD) and fiber-optic gyroscope (FOG), the tracking performance (TP) and anti-disturbance ability (ADA) characterized by boresight error are of equal importance. Generally, the position tracking loop, limited by the image integration time of CCD, would be subject to a non-negligible delay and low-sampling rate, which could not minimize the boresight error. Although the FOG-based velocity loop could enhance the ADA of the system, it is still insufficient in the case of some uncertain disturbances. In this paper, a feedforward control method based on the results of error and disturbance observation was proposed. The error observer (EOB) based on the CCD data and model output essentially combined the low-frequency tracking feedforward and closed-loop disturbance observer (DOB), which could simultaneously enhance the low-frequency TP and ADA. In addition, in view of the poor low-frequency performance of the FOG due to drift and noise that may result in the inaccuracy of the observed low-frequency disturbance, the FOG-based DOB was used to improve the relatively high-frequency ADA. The proposed method could make EOB and DOB complementary and help to obtain a high-precision MOTS, for in practical engineering, we give more attention to the low-frequency TP and full-band ADA. Simulations and experiments demonstrated that the proposed method was valid and had a much better performance than the traditional velocity and position double-loop control (VPDC).

Keywords: feedforward control; mobile optoelectronic tracking system; error observer; disturbance observer; tracking performance; anti-disturbance ability; model reference

---

1. Introduction

The charge-coupled device (CCD)-based mobile optoelectronic tracking system (MOTS), commonly mounted on vehicles, ships, airplanes and satellites, is mainly used for astronomical observation, free space communication, searching and target tracking [1–5]. The closed-loop control of the system is based on the boresight error detected by a CCD. The value of boresight error could reflect the tracking performance (TP) and anti-disturbance ability (ADA), both of which are equally important in a moving platform being full of various disturbances. Due to the image integration time of the CCD, the position tracking loop would be subject to a non-negligible delay and low-sampling rate [6,7], which are major causes of instability and performance deterioration [8–10]. Scholars have adopted many optimization methods to enhance the tracking accuracy and decrease the bad influence of the delay. A PID-I method, with an integration added to the controller, was proposed to reduce the steady-state error of the system [11], which, however, would affect the stability of the system and decrease its dynamic performance. In order to eliminate the effect of delay on system stability, a Smith predictor was introduced to the tracking loop [12,13]; however, the delay was moved out from the closed loop but still existed in the system, which would restrict the TP. As reported, a multi-loop control structure based on MEMS inertial sensors could increase the bandwidth of the system [14]. Nevertheless, there was little low-frequency improvement of TP. Compared to the difficulties in enhancing the TP, the ADA of the system is relatively easier to improve, for the disturbances usually originate from the base which could be measured by inertial sensors with little delay. Therefore, an inertial sensor, such as a fiber-optic gyroscope (FOG) mounted parallel to the boresight, is commonly used to establish a high-rate inner loop, which would increase the whole ADA of the system [15,16]. However, the ADA of the closed-loop control is still insufficient for plant uncertainties and large-magnitude disturbances. In summary, the feedback control method is limited in improving either the TP or ADA of the system.

In order to get a satisfactory performance, it is necessary to perform feedforward control, including the tracking and disturbance feedforward. Theoretically, the errors could be significantly reduced or even eliminated, and almost all measurable disturbances could be suppressed. Unfortunately, it was difficult to get the trajectory of the target and extract the disturbance signals. A predictive tracking method combining the boresight error with angular sensor for synthesizing the target trajectory was proposed to compensate the errors caused by time delay [17,18]. However, an additional position sensor was required, which was only suitable for the condition with low measurement noise. Similarly, in order to detect the disturbances, additional sensors should be equipped on the pedestal [19,20]; the wind disturbance and cogging force cannot be reflected from the pedestal. Hence, the feedforward control based on direct measurement could not be easily implemented, especially for the space and cost limited occasions.

In this paper, an unconventional feedforward control method based on the error and disturbance observation was proposed considering the velocity and position double-loop control (VPDC). The CCD-based error observer (EOB) combined the differential of the boresight error and model output with a delay to generate a composite velocity, which simultaneously contained the delayed signals of the target motion and disturbance. Since the delay had little effect on the very low-frequency signal, the low-frequency items of the composite velocity could be fed forward to the velocity closed loop. The EOB was equivalent to a coalition of the low-frequency tracking and disturbance feedforward. Although the TP improvement was in the low frequency, it was satisfactory because the target motion signal mainly distributed there. Unlike the TP, the ADA improvement only in low frequency brought by EOB was not enough because external disturbances nearly distributed in the full frequency band. Therefore, a FOG-based disturbance observer (DOB) was continuously added to the inner loop to increase the ADA. Unlike the direct disturbance feedforward method, by which additional sensors should be equipped on the pedestal to extract disturbance, DOB could acquire the disturbance through the difference between the data of the existing sensors and the model output [21,22]. Since the FOG's low-frequency signal was weak and susceptible to drift and noise, which resulted in the inaccuracy of the observed low-frequency disturbance, the additional DOB mainly benefited the high-frequency ADA. Hence, the proposed method could make the CCD-based EOB and FOG-based DOB complementary. To verify the effectiveness of the method, a platform consisting of groups of the fast steering mirror system was established, which was the core component of MOTS [14]. Experiments demonstrated that the proposed system had a good TP in low frequency and a strong ADA in a wide band.

This paper is organized as follows. Section 2 introduces the physical structure of the MOTS, the basic VPDC method and the common tracking and disturbance feedforward way based on direct measurement. Section 3 analyzes the proposed feedforward method based on the EOB and DOB, which could provide theoretical derivations. Section 4.1 discusses the matter of how to design the EOB controller Q_1 to maximize the performance of the EOB under the condition of guaranteed gain and phase margin. Section 4.2 focuses on the design of the DOB controller Q_2 and analyzes the promotion of ADA. Section 5 is the experimental part, indicating the detailed improvement of the TP and ADA by the proposed method. Section 6 lists the concluding remarks.

2. The CCD-Based MOTS with Traditional Control Methods

The basic configuration of the MOTS is shown in Figure 1. The light from the target could pass through the reflective surface of the rotating mirror, which would then be detected by CCD to calculate the boresight error. After receiving the boresight error, the controller would drive the voice motors to make the mirror rotate accordingly, thereby tracking the target and resisting the impact of the external disturbances. To enhance the ADA, a FOG was mounted on the lens barrel to measure the angle velocity. In addition, the velocity closed loop would increase the stiffness of the object and make it easier to control the platform.

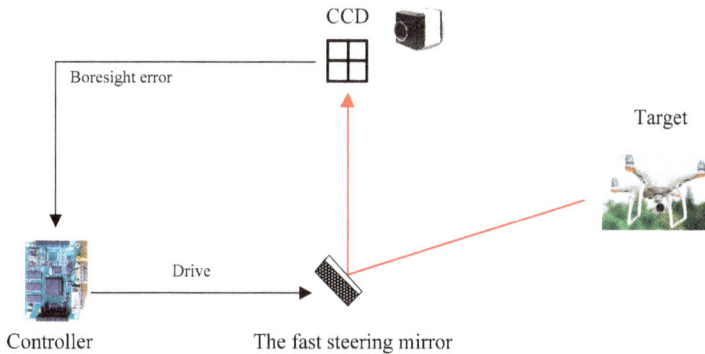

Figure 1. Configuration of the MOTS.

2.1. The Basic VPDC Control Method

In Figure 2, the basic control structure of the VPDC was presented. The error transfer functions of the tracking and disturbance were respectively S_R and S_D, as shown in Equations (1) and (2).

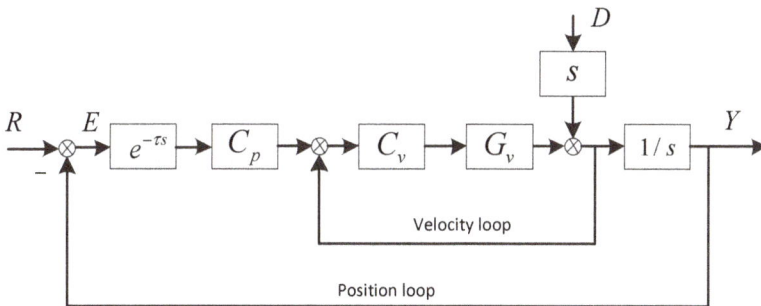

Figure 2. The VPDC structure. G_v refers to the velocity open-loop transfer function. C_v and C_p respectively refer to the velocity controller and position controller. $e^{-\tau s}$ refers to the delay element of the CCD. E refers to the boresight error without delay. R and D respectively refer to the target signal and the external disturbance signal. Y refers to the corresponding movement of the boresight.

$$S_R = \frac{E}{R} = \frac{1}{1 + C_p \frac{C_v G_v}{1 + C_v G_v} \frac{1}{s} e^{-\tau s}} \approx \frac{1}{1 + C_p \frac{1}{s} e^{-\tau s}} \tag{1}$$

$$S_D = \frac{E}{D} = \frac{1}{1+C_vG_v+C_pC_vG_v\frac{1}{s}e^{-\tau s}} = \frac{1}{1+C_vG_v} \cdot \frac{1}{1+C_p\frac{C_vG_v}{1+C_vG_v}\frac{1}{s}e^{-\tau s}}$$
$$\approx \frac{1}{1+C_vG_v} \cdot \frac{1}{1+C_p\frac{1}{s}e^{-\tau s}} \tag{2}$$

where $C_vG_v/(1+C_vG_v) \approx 1$ in low frequency, because the high-sampling rate velocity loop commonly has a high bandwidth over 100 Hz [14]. As we all know, the smaller the error transfer function is, the higher the accuracy will be. From Equations (1) and (2), it could be concluded that the TP could be slightly improved by the velocity loop. However, the velocity loop could significantly improve the ADA and the enhanced part is $|1/(1+C_vG_v)|$. Unfortunately, when tracking a high-velocity target with strong external disturbance from the pedestal, the TP and ADA would still be insufficient. If the structure was not modified, we could only increase the gain C_p and C_v or add more integral elements to enhance the performance. However, these would decrease the margin and could even make the system unstable. To get a high-precision system under complex conditions, the feedforward branch should be introduced.

2.2. The Conventional Feedforward Based on Direct Measurement

The feedforward control, as a robust method in industrial control, can effectively decrease the influence of delay and establish a high real-time and high-precision system. The VPDC-based direct feedforward structure is shown below (Figure 3).

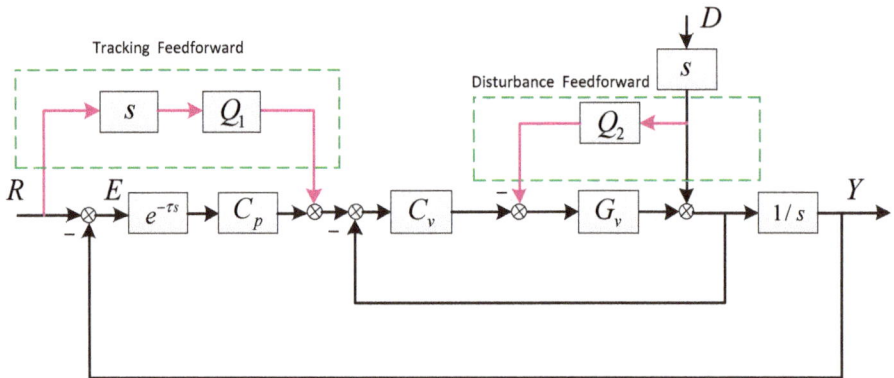

Figure 3. The direct feedforward structure based on VPDC.

The error transfer functions are as follows.

$$S'_R = \frac{E}{R} = \frac{1-Q_1 \cdot \frac{C_vG_v}{1+C_vG_v}}{1+C_p\frac{C_vG_v}{1+C_vG_v}\frac{1}{s}e^{-\tau s}} \approx \frac{1-Q_1}{1+C_p\frac{1}{s}e^{-\tau s}} \tag{3}$$

$$S'_D = \frac{E}{D} = \frac{1-Q_2\cdot G_v}{1+C_vG_v+C_pC_vG_v\frac{1}{s}e^{-\tau s}} = \frac{1}{1+C_vG_v} \cdot \frac{1-Q_2\cdot G_v}{1+C_p\frac{C_vG_v}{1+C_vG_v}\frac{1}{s}e^{-\tau s}}$$
$$\approx \frac{1}{1+C_vG_v} \cdot \frac{1-Q_2\cdot G_v}{1+C_p\frac{1}{s}e^{-\tau s}} \tag{4}$$

Compared to Equations (1) and (2), if Q_1 and Q_2 are designed properly, the error could be reduced to 0, theoretically. However, in fact, the pure feedforward control is a kind of open-loop control highly relying on the object model. Since the mathematical model could only be built accurately at low and middle frequencies, the promotion mainly concentrated in these bands. In addition, it was difficult to get signals of the target movement and disturbance. Firstly, there is no sensor that could directly detect the motion state of the target. If the sensors' fusion method is adopted to predict the target trajectory, an additional sensor should be used to measure the position of the platform. The prediction method requires lots of computation and is only suitable for low noise environments. Secondly, in order to extract the external disturbance from the pedestal, additional sensors are also required. Moreover, it is difficult to identify the disturbance source and more auxiliary equipments are required. Therefore, the feedforward based on direct measurement is also inappropriate in engineering.

3. The EOB and DOB-Based Indirect Feedforward Control

3.1. The CCD-Based EOB

The proposed way of combining EOB and DOB could be regarded as an indirect feedforward control of model reference. The EOB structure is shown in Figure 4. The inner velocity loop has changed the velocity transfer function and it could be treated as 1 in low frequency. The given velocity v_{ref} after passing a delay element is actually an output of the inner closed-loop model. Through combining the differential of the boresight error and the model output, a composite velocity is produced, which simultaneously contains the information of the target and disturbances. In order to better understand the essence of the EOB, its equivalent structure, as shown in Figure 5, should be referred to.

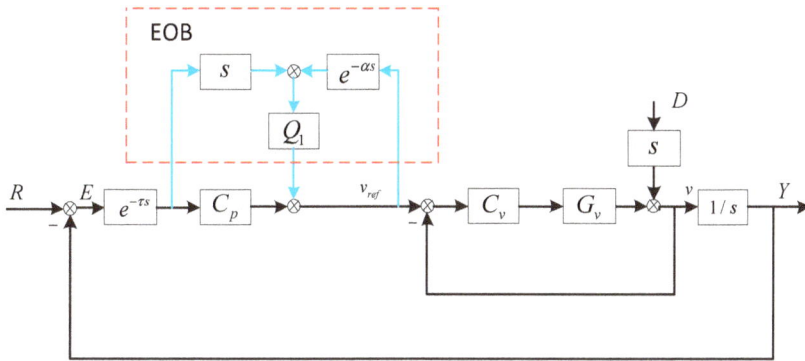

Figure 4. The EOB structure based on VDPC. e^{-as} refers to an artificially added delay and $\alpha = \tau$. Q_1 refers to the EOB controller.

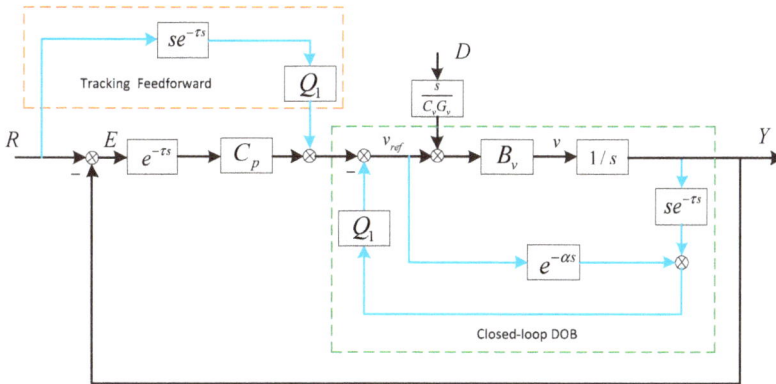

Figure 5. The equivalent structure of EOB. $B_v = C_v G_v / (1 + C_v G_v)$ refers to the closed-loop transfer function. $B_v \approx 1$ at a low frequency.

In Figure 5, the EOB is split into a tracking feedforward structure and a closed-loop DOB. In the tracking feedforward, the input of Q_1 refers to the target's velocity with the delay of CCD. Similarly, the input of Q_1 in DOB refers to an external disturbance with a same delay. These delayed signals are valuable, which can be fed forward to the system at a low frequency, because the phase lag caused by delay would be little at the very low frequency. However, at the high frequency, the phase lag would be great, which could lead to instability of the system. Hence, to reduce the negative influence of the delayed input, the low-pass filter Q_1 should be adopted, which would only allow the passing of low-frequency signals. Of course, because of the existing of the delay, the effect of EOB is slightly worse than the direct feedforward way based on sensors fusion, but this still can be accepted. Therefore, EOB is an incomplete feedforward way working at the low frequency.

To get the error transfer functions easier, the EOB in Figure 4 can also be regarded as an equivalent position controller as follows.

$$C'_p = \frac{sQ_1 + C_p}{1 - Q_1 e^{-\alpha s}} \tag{5}$$

With C'_p, the error transfer functions in Figure 4 can be derived below.

$$
\begin{aligned}
\widehat{S}_R = \frac{E}{R} &= \frac{1 - Q_1 \cdot e^{-\alpha s}}{1 + Q_1(\frac{C_p G_p}{1+C_v G_v} e^{-\tau s} - e^{-\alpha s}) + C_p \frac{C_v G_v}{1+C_v G_v} \frac{1}{s} e^{-\tau s}} \\
&\approx \frac{1 - Q_1 \cdot e^{-\alpha s}}{1 + C_p \frac{1}{s} e^{-\tau s}}
\end{aligned}
\tag{6}
$$

$$
\begin{aligned}
\widehat{S}_D = \frac{E}{D} &= \frac{1}{1+C_v G_v} \cdot \frac{1}{1 + \frac{sQ_1 + C_p}{1 - Q_1 e^{-\tau s}} \frac{C_v G_v}{1+C_v G_v} \frac{1}{s} e^{-\tau s}} \\
&= \frac{1}{1+C_v G_v} \cdot \frac{1 - Q_1 e^{-\alpha s}}{1 + Q_1(\frac{C_p G_p}{1+C_v G_v} e^{-\tau s} - e^{-\alpha s}) + C_p \frac{C_v G_v}{1+C_v G_v} \frac{1}{s} e^{-\tau s}} \\
&\approx \frac{1}{1+C_v G_v} \cdot \frac{1 - Q_1 e^{-\alpha s}}{1 + C_p \frac{1}{s} e^{-\tau s}}
\end{aligned}
\tag{7}
$$

Compared to Equations (3) and (4), the values of Equations (6) and (7) could not reach 0 due to the existence of delay. However, it could be close to 0 at the very low frequency, because $e^{-\alpha s} \approx 1$ under this condition. Theoretically, the lower bandwidth of Q_1 could benefit the stability of the system, but the very low frequency would reduce the effect of feedforward. In order to maximize the benefits of the feedforward while ensuring the stability, the matter of how to choose Q_1 with guaranteed gain margin (GM) and phase margin (PM) will be discussed in Section 4.

3.2. The Additional FOG-Based DOB

EOB can apparently enhance the low-frequency TP and ADA. It is adequate to complete the tracking because the main frequencies of the target signal are low. However, the frequencies of external disturbance distribute in a wide band. In addition to the low-frequency sway, the MOTS could also be greatly affected by relatively high-frequency mechanical vibration and electromagnetic interference. Therefore, to further improve the ADA and release the potential of the FOG only used in feedback, the FOG-based DOB was added into the velocity loop. FOG had a high bandwidth over 100 Hz, while at the low frequency, its signal was susceptible to drift and noise. Hence, the DOB could extract an accurate disturbance at the relatively high frequency, which can promote the ADA a lot in the higher band. The structure combining EOB with DOB is presented in Figure 6.

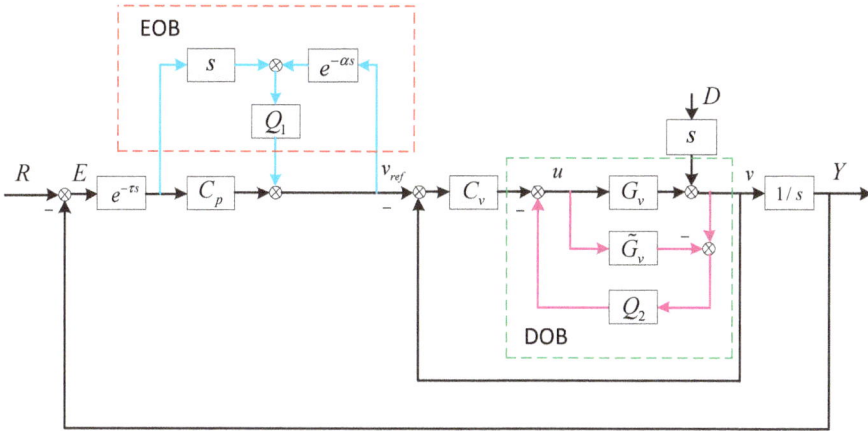

Figure 6. The VPDC structure enhanced by EOB and DOB. $\widetilde{G}_v$ refers to the approximate velocity model of the platform and $\widetilde{G}_v \approx G_v$. Q_2 refers to the DOB controller.

The closed-loop velocity is given as follows

$$v = uG_v + sD \tag{8}$$

$$u = (v_{ref} - v)C_v - (v - u\widetilde{G}_v)Q_2 \tag{9}$$

After calculation,

$$v = \frac{C_vG_v}{1 + C_vG_v + Q_2(G_v - \widetilde{G}_v)}v_{ref} + \frac{(1 - Q_2\widetilde{G}_v)s}{1 + C_vG_v + Q_2(G_v - \widetilde{G}_v)}D \tag{10}$$

Since $E = R - \frac{1}{s}v$ and $v_{ref} = E \cdot e^{-\tau s}C'_p$, we could get

$$
\begin{aligned}
\hat{S}_R = \frac{E}{R} &= \frac{1 - Q_1e^{-as}}{1 - Q_1e^{-as} + \frac{C_vG_v}{1 + C_vG_v + Q_2(G_v - \widetilde{G}_v)}(C_p + sQ_1)e^{-\tau s}\frac{1}{s}} \\
&\approx \frac{1 - Q_1e^{-as}}{1 - Q_1e^{-as} + (C_p + sQ_1)e^{-\tau s}\frac{1}{s}} \\
&\approx \frac{1 - Q_1e^{-as}}{1 + C_pe^{-\tau s}\frac{1}{s}}
\end{aligned}
\tag{11}
$$

$$
\begin{aligned}
\hat{S}_D = \frac{E}{D} &= \frac{(1 - Q_2\widetilde{G}_v)}{1 + C_vG_v + Q_2(G_v - \widetilde{G}_v)} \cdot \frac{(1 - Q_1e^{-as})}{1 - Q_1e^{-as} + \frac{C_vG_v}{1 + C_vG_v + Q_2(G_v - \widetilde{G}_v)}(C_p + sQ_1)e^{-\tau s}\frac{1}{s}} \\
&\approx \frac{(1 - Q_2\widetilde{G}_v)}{1 + C_vG_v + Q_2(G_v - \widetilde{G}_v)} \cdot \frac{(1 - Q_1e^{-as})}{1 - Q_1e^{-as} + (C_p + sQ_1)e^{-\tau s}\frac{1}{s}} \\
&\approx \frac{(1 - Q_2\widetilde{G}_v)}{1 + C_vG_v} \cdot \frac{(1 - Q_1e^{-as})}{1 + C_pe^{-\tau s}\frac{1}{s}}
\end{aligned}
\tag{12}
$$

Equation (11) is approximately equal to Equation (6), which means that the FOG-based DOB could not enhance the TP, but slightly affect the stability of the system. Comparing Equation (12) with Equation (7), it could be obviously found that the ADA was continuously improved. Then, at the low frequency $1 - Q_1e^{-as}$ was close to 0 and at the relatively high frequency $1 - Q_2\widetilde{G}_v$ was close to 0, which means that the CCD-based EOB and FOG-based DOB could complement each other and a system with strong ADA in a wide band was acquired. The design of Q_1 and Q_2 will be discussed in the following section.

4. Parameters Design and Performance Analysis

4.1. The Design of Q_1 and the Performance Improvement with EOB

Firstly, we should determine the form of the original position controller C_p in Figure 2. According to the previous analysis, the inner loop as the controlled object was close to 1. It is easy to verify whether $C_p = k = \frac{\pi}{4\tau}$ can stabilize the platform with PM more than 45° and GM more than 6 dB. The crossover frequency and gain frequency of the outer loop can be defined as ω_c and ω_g. We can easily get $\omega_c = \frac{\pi}{4\tau}$ and $\omega_g = \frac{\pi}{2\tau}$. To investigate the stability of the system, the open-loop transfer function was presented as follows.

$$G_{open} = C_p \frac{1}{s} e^{-\tau s} \tag{13}$$

After adding the EOB, the open-loop transfer can be changed to Equation (14).

$$
\begin{aligned}
G'_{open} &= C'_p \frac{1}{s} e^{-\tau s} \\
&= \frac{sQ_1 C_p^{-1} + 1}{1 - Q_1 e^{-\alpha s}} \cdot \left(C_p \frac{1}{s} e^{-\tau s} \right)
\end{aligned}
\tag{14}
$$

In Equation (14), the low-pass filter Q_1 may be set as a simple first-order form as follows.

$$Q_1 = \frac{1}{1 + Ts} \tag{15}$$

To make the equivalent controller C'_p have no effect on the stability margin, two restrictions should be met: (1) $\arg[L(j\omega_c)] \geq 0$; (2) $-20\log[L(j\omega_g)] \geq 0$, where $L(s)$ should be set as follows.

$$L(s) = \frac{sQ_1 C_p^{-1} + 1}{1 - Q_1 e^{-\alpha s}} = \frac{ks + (Ts + 1)}{(Ts + 1) - e^{-\alpha s}} \tag{16}$$

According to Appendix A, we know that the phase and gain loss of $L(s)$ is inevitable. In order to meet the stability condition of $PM > 45°$ and $GM > 6$dB, we should reduce the controller gain appropriately. Although it would cause the close-loop bandwidth to drop slightly, it is acceptable because the low-frequency ability of suppressing errors is more important than the bandwidth. With the EOB structure, the previous controller C_p is substituted by $0.8C_p$, accompanied with $T = 6\alpha$ and $\alpha = \tau$.

Comparing Equations (6) and (7) with Equations (1) and (2), it can be found that the promotion of the closed-loop performance depends on the amplitude of $1 - Q_1 e^{-\alpha s}$.

$$\left| 1 - Q_1 e^{-j\tau\omega} \right|^2 = 1 + \frac{1 + 2T\omega \sin(\tau\omega) - 2\cos(\tau\omega)}{1 + (T\omega)^2} \tag{17}$$

Obviously, if $\omega \to 0$, then $\left| 1 - Q_1 e^{-j\tau\omega} \right|^2 \to 0$, which means that the improvement is huge at the very low frequency. Since $1 + 2T\omega \sin(\tau\omega)$ always increases and $2\cos(\tau\omega)$ decreases when $\tau\omega$ changes from 0 to 0.5π, it is evident that $\omega_o \in (0, \omega_g)$ exists, which makes $1 + 2T\omega_o \sin(\tau\omega_o) = 2\cos(\tau\omega_o)$. When $\omega < \omega_o$, $\left| 1 - Q_1 e^{-j\tau\omega} \right|^2 < 1$ the system performance would be enhanced.

The following discussion is on the amplification of $\left| 1 - Q_1 e^{-j\tau\omega} \right|^2$ for errors. Equation (17) can be rewritten as follows.

$$\left| 1 - Q_1 e^{-j\tau\omega} \right|^2 = 1 + \frac{1 + 2T\omega \sin(\tau\omega) - 2\cos(\tau\omega)}{1 + (T\omega)^2} \leq 1 + \frac{1 + 2T\omega}{1 + (T\omega)^2} < 3 \tag{18}$$

It means that the amplification of the errors at the medium frequency is limited. Moreover, actually, $\left| 1 - Q_1 e^{-j\tau\omega} \right|^2$ cannot reach the boundary value in Equation (18) and approaches 0 when ω is big enough.

In this paper, the CCD's delay is 0.02 s (two frames 100 Hz sampling rate). Substitute all the parameters to $1 - Q_1 e^{-as}$ and the Simulation is shown in Figure 7. The performance improvement was big at the low frequency and even reached -20 dB at 0.1 Hz. The peak in the middle frequency from 2 Hz to 20 Hz was consistent with the previous analysis, but it was very small and not at the low frequency for tracking. On the contrary, for suppressing disturbances, the relatively high-frequency performance cannot be ignored. Hence, the FOG-based DOB should be added.

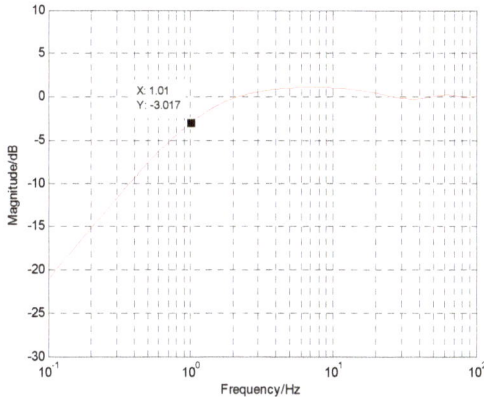

Figure 7. The simulation of the performance promotion with EBO.

4.2. The Design of Q_2 and the ADA Improvement with DOB

The FOG-based DOB method depends on the model of the platform. The fine tracking model of the MOTS is as follows, which contains a differential element, a mechanical resonance element with natural frequency ω_n, and an inertial element with electrical time constant T_e. Then, with the spectrum fitting method in system identification, the detailed parameters could be determined. Commonly, ω_n is several Hz and $T_e \ll 1$.

$$G_v = \frac{Ks}{\frac{s^2}{\omega_n^2} + \frac{2\xi}{\omega_n}s + 1} \cdot \frac{1}{T_e s + 1} \tag{19}$$

Before adding the DOB structure into the inner loop, we should complete the design of the velocity closed loop. To get a high-bandwidth inner loop with a sufficient PM and GM, the controller can be set as follows, which would make the open-loop transfer function as an approximate integration element. The anti-resonant link can be used to compensate the resonance in G_v, $T_e s + 1$ to promote the phase lag, and the inertial element can be used to filter the high-frequency noise.

$$C_v = \frac{k_v \left(\frac{s^2}{\omega_n^2} + \frac{2\xi}{\omega_n}s + 1 \right)}{s^2} \cdot \frac{(T_e s + 1)}{T_v s + 1} \tag{20}$$

After establishing the velocity closed loop, the position open-loop transfer function could still be an approximate integration element and the position controller C_p should be set as a proportional link according to Section 4.1.

After completing the VPDC design, the DOB could be introduced without affecting the stability of the system. According to Figure 6 and Equation (12), the ideal DOB controller Q_2 should be the inverse of G_v, as shown below,

$$Q_2 = (G_v)^{-1} = \frac{\left(\frac{s^2}{\omega_n^2} + \frac{2\xi}{\omega_n}s + 1 \right)(T_e s + 1)}{Ks} \tag{21}$$

In Equation (21), the order of the numerator order is higher than that of the denominator, which cannot be accomplished in physics. In addition, as the FOG's signal is commonly accompanied with drift at the low frequency, the integration in Q_2 would exacerbate this situation and result in the saturation of the driver, which would affect the stability. To solve the problem, the controller Q_2 should make a compromise, although this would sacrifice some performances at the very low and high frequency. The practical controller is shown as follows.

$$Q_2 = \frac{\frac{s^2}{\omega_n^2} + \frac{2\zeta}{\omega_n}s + 1}{K(s+b)(T_1s+1)} \tag{22}$$

In Q_2, the integral is changed to be $1/(s+b)$, making the integral effect disappear at the low frequency. What's more, an inertial element, in which $T_1 \ll 1$, is used to filter the high-frequency signals and abandon the compensation for high frequency disturbances because the model of the platform is commonly not accurate at the very high frequency. From Equation (12), the ADA promotion from DOB is due to $1 - Q_2\tilde{G}_v$, and now attention is attached to its value.

$$
\begin{aligned}
1 - Q_2\tilde{G}_v &= 1 - \frac{\frac{s^2}{\omega_n^2} + \frac{2\zeta}{\omega_n}s + 1}{K(s+b)(T_1s+1)} \cdot \frac{Ks}{\frac{s^2}{\omega_n^2} + \frac{2\zeta}{\omega_n}s + 1} \cdot \frac{1}{T_es+1} \\
&= 1 - \frac{s}{(s+b)(T_1s+1)(T_es+1)} \\
&= \frac{T_1T_es^3 + (T_1+T_e+bT_1T_e)s^2 + (bT_1+bT_e)s + b}{T_1T_es^3 + (T_1+T_e+bT_1T_e)s^2 + (1+bT_1+bT_e)s + b}
\end{aligned}
\tag{23}
$$

At the very low frequency, $1 - Q_2\tilde{G}_v \approx b/b = 1$, there is no improvement for ADA. At the medium frequency, $1 - Q_2\tilde{G}_v \approx b/(s+b)$, there is a maximum -20 dB promotion and as ω grows, the promotion is smaller. At the very high frequency, $1 - Q_2\tilde{G}_v \approx (T_1T_es^3/T_1T_es^3) \approx 1$, the lift disappears again. From the analysis, $1 - Q_2\tilde{G}_v$ is a band-pass filter, which does not work at very low or high frequency. However, this is acceptable because the EOB could enhance the low-frequency performance and the very high-frequency ADA commonly relies on the system's mechanical design. The ADA improvement of DOB was simulated in Figure 8, and the whole effect brought by both EOB and DOB was also presented. The result was as expected. Like a notch filter, the pure DOB can mainly function at the medium frequency, while the effect of EOB and DOB can benefit both the low- and medium-frequency performance.

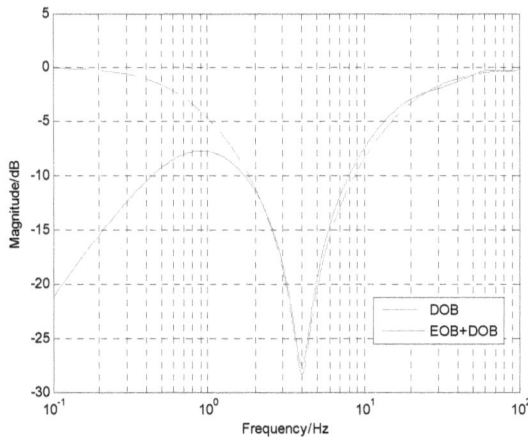

Figure 8. The simulation of the ADA improvement.

5. Experimental Verification

The experimental setup was presented in Figure 9, which involved three fast steering mirror systems used in the disturbance isolation table. One was the controlled object named Tracking Mirror, one was Target Mirror for simulating target motion, and one was Disturbance Mirror for simulating external disturbances. The Tracking Mirror was fixed above the Disturbance Mirror. Since the fast steering mirror was a symmetrical two-axis system, we only need to pay attention to single-axis motion. The laser could emit light as a visual axis reference. Then, the light could be reflected by the Target Mirror and entered into the phase sensitive demodulator (PSD), a substitute for CCD. We could control the motion of Target Mirror and Disturbance Mirror to simulate the target's motion and the external interference. The controller could receive the boresight error from PSD and the platform's velocity detected by a FOG, to stabilize the boresight. The PSD could run in at the sampling rate of 100 Hz with an artificially added delay to imitate the CCD. The FOG could run at the rate of 5000 Hz.

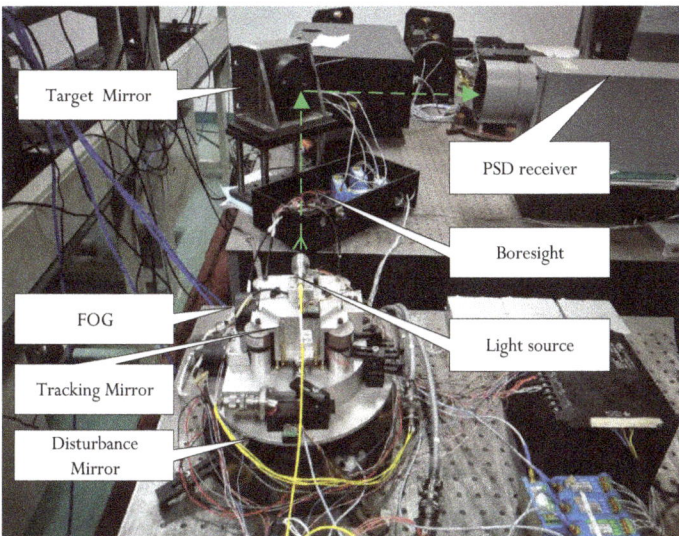

Figure 9. Experimental setup.

Before designing the closed-loop controller, the model of the platform should be acquired with a spectrum fitting method. In order to identify the parameters in Equation (19), the driver should output a sinusoidal signal of varying frequency to actuate the Tracking FSM. Then the FOG would be used to detect the motion state; comparing the output to the input, the open-loop bode response of the velocity can be drawn with the blue line, as shown in Figure 10. Finally, the parameters were adjusted to make the red curve of the model coincide with the blue one, and a high-precision transfer function could be acquired, as shown in Equation (24).

$$\tilde{G}_v(s) = \frac{2.3s}{0.00072s^2 + 0.0202s + 1} \cdot \frac{1}{0.0005s + 1} \tag{24}$$

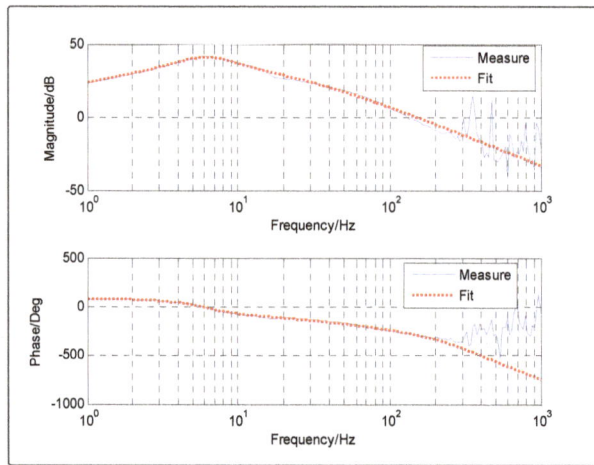

Figure 10. The open-loop bode response of the velocity.

The closed-loop bode response of the velocity was exhibited in Figure 11, in which the bandwidth passed 100 Hz, and the resonance peak of the open-loop velocity model under 7 Hz was eliminated, benefiting the controller design of the outer position loop. Below 10 Hz, the amplitude was close to 0 dB and the phase loss was less than 6°, which indicated that the closed-loop transfer function could be regarded as 1 in this low frequency band. Therefore, it was enough to take a proportional link as the position controller.

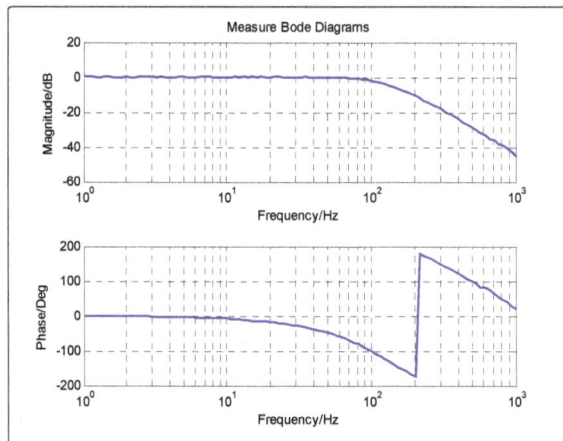

Figure 11. The closed-loop bode response of the velocity.

5.1. The Improvement of the TP with EOB

In Figure 12, the closed-loop bode responses of the position with or without EOB were described. When the EOB was introduced, the amplitude of the position controller should drop a little to guarantee enough margins. Although the bandwidth of the system with EOB was decreased by about 2 Hz, it would make no difference, because the performance below 1 Hz was adopted for tracking.

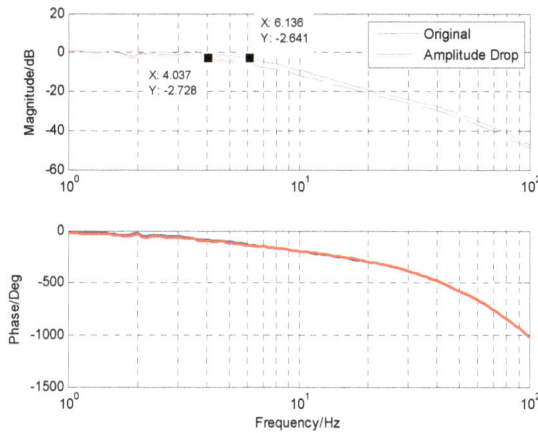

Figure 12. The closed-loop bode responses of the position.

The error suppression bode response of the MTOS was presented in Figure 13, which only provided the target signal. Three methods were listed overall, the basic VPDC method, the direct feedforward method and the EBO method based on the VPDC. Obviously, the direct feedforward method based on sensors fusion with the predictive way had the best error suppression performance. However, the performance of the EBO method was close to the direct feedforward, which signified that the previous analysis was right and the proposed way was approximately equivalent to the direct feedforward method. Moreover, compared with the direct feedforward, the EBO method did not need an additional position sensor to measure the angular of the platform and cost a smaller amount of computation. Below 2 Hz, the system with EOB would have a better ability of error suppression than the pure VPDC without EOB; actually, it would be valid for both tracking and anti-disturbance. With the decrease of the frequency, the promotion of ability would be more apparent and could even reach −20 dB under 0.1 Hz. In the frequency band between 2 Hz and 18 Hz, the performance of the pure VPDC was a little better than the system with EBO, as shown in Figure 7, and the performance could be accepted. In Figure 14, the time-domain residual error was presented with the given target signal of different frequencies. In the frequency domain below 1 Hz, the EBO can make a big difference.

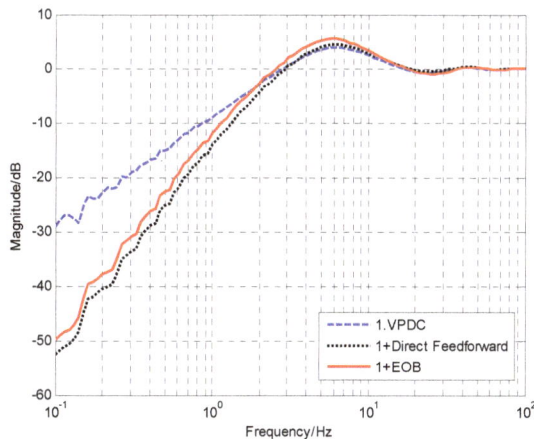

Figure 13. The error suppression response of the MOTS.

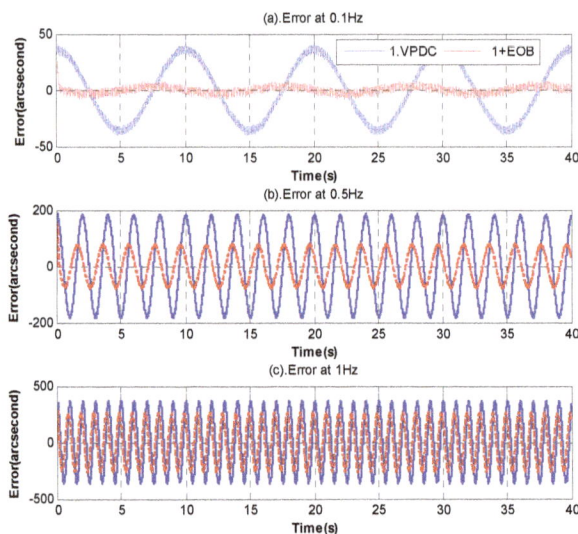

Figure 14. The residual tracking errors in different frequencies.

5.2. The Improvement of the ADA with the Combination of EOB and DOB

According to Equation (22), the DOB controller Q_2 is not the ideal value and a concession should be made. Parameter b can determine the starting point of disturbance compensation. Considering that the used FOG in this experiment had a poor performance below about 1 Hz, the actually used Q_2 was set as Equation (25) and all the parameters of controllers were listed in Table 1.

$$Q_2 = \frac{0.00072s^2 + 0.0202s + 1}{2.3(s+5)} \cdot \frac{1}{0.0065s + 1} \tag{25}$$

Table 1. The parameters of controllers. When EOB is added, C_p should be changed to $C'_p = 0.8C_p$ for sufficient PM and GM according to the previous analysis.

Controllers	C_p	C_v	Q_1	Q_2
Parameters	$\frac{\pi}{4\tau} = 39.25$	$k_v = 16.8$ $T_e = 0.0005$ $T_v = 0.0003$ $\omega_n = 5.9\text{Hz}$ $\zeta = 0.3764$	$T = 0.12$	$K = 2.3$ $b = 5$ $T_1 = 0.0065$ $\omega_n = 5.9\text{Hz}$ $\zeta = 0.3764$

The disturbance suppression bode response is presented in Figure 15, which involved four situations in all, showing the pure VPDC structure and the various combinations of EOB and DOB based on VPDC; the VPDC without any feedforward had a relatively poor ADA, especially in the middle frequency domain of about 5 Hz, which approached 0 dB. The EOB has apparently promoted the ADA of the low frequency below 1 Hz, sometimes (for example, on a ship) there would be a shake of carrier with a strong amplitude. The reason why the promotion was restricted at the frequency of 0.1 Hz was that the performance of the system almost reached the limitation under this noise condition. The DOB mainly worked at the medium frequency, where there was a maximum increase of -20 dB, because the measured signals and the reference model were relatively more accurate in this band. With the simultaneous help of EOB and DOB, the ADA had an improvement (at least -10 dB) in both the low and medium frequencies, which means that the system's precision would nearly be increased by an order of magnitude. The proposed method could not enhance the very high-frequency ADA, which mainly depended on the mechanical design. Figure 16 shows the time-domain residual error with a given disturbance signal of different frequencies. Obviously, with the proposed way, the residual error was reduced below 30 Hz.

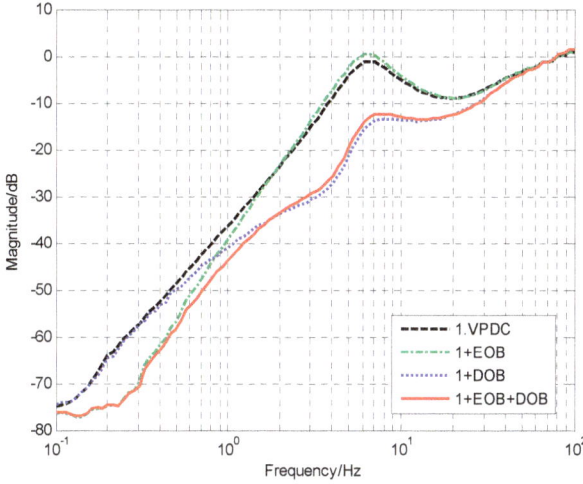

Figure 15. The disturbance suppression response of the MOTS.

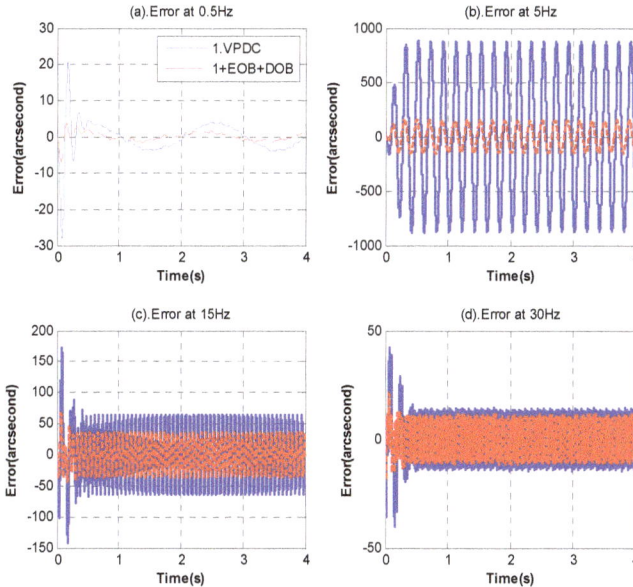

Figure 16. The residual stabilization errors in different frequencies.

In order to verify the level of accuracy under the actual engineering conditions, we should detect the residual errors when simultaneously giving the target signal and disturbances to the system. In Figure 17, the results were presented and there was huge promotion. The target signal was 0.28° 0.2 Hz and the disturbances consisted of 0.28° 0.5 Hz, 0.0028° 5 Hz and 0.028° 50 Hz. The RMS error of the pure VPDC was 8.6504″ and the RMS error with additional EOB and DOB was 1.8111″, signifying that the proposed way of observation was valid.

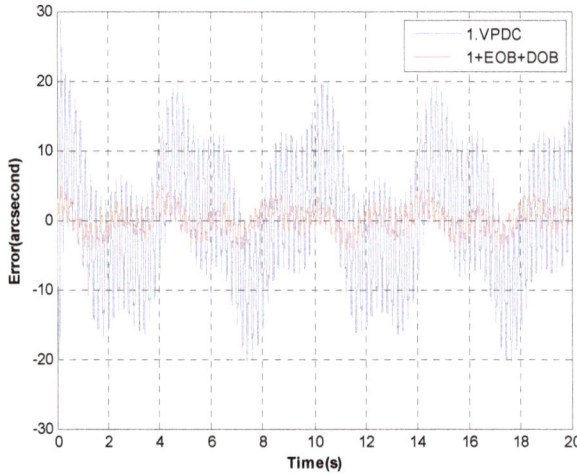

Figure 17. The residual error with the given signals of the target and disturbances.

6. Conclusions

In this paper, an unconventional feedforward method was introduced based on indirect measurement. EOB and DOB were combined to enhance the low-frequency TP and improve the ADA in a wide frequency band. The CCD and model output-based EOB were essentially a combination of the incomplete tracking and disturbance feedforward, which increases the low-frequency performance. The FOG-based DOB was a supplement for enhancing the relatively high-frequency ADA. The potential of the sensors was fully released and the low-bandwidth CCD and high-bandwidth FOG were made complementary. The multi-loop feedback control and feedforward control were simultaneously adopted to promote the accuracy and stability of the system. The design of Q_1 and Q_2 was analyzed, which is easy to be implemented in engineering. Experimental results demonstrated that a high-performance MOTS with a satisfied TP and strong ADA was acquired.

Since the performance of the system enhanced by the proposed method has approached to the limitation in this noise condition, we will consider regarding noise as an input and adopt some time-domain filters to decrease the impact of noise on the system.

Author Contributions: Conceptualization, Y.L., Y.H. and Y.M.; Data curation, Y.M.; Formal analysis, Y.L.; Funding acquisition, Y.M.; Investigation, W.R.; Methodology, Y.L. and Q.W.; Project administration, X.Z.; Resources, Y.H.; Software, W.R. and Q.H.; Supervision, Q.H.; Validation, X.Z.; Visualization, W.R.; Writing—original draft, Y.L.; Writing—review & editing, Y.L.

Funding: This research received no external funding.

Acknowledgments: We would extend our sincere gratitude to the Chinese Academy of Science for her sponsor.

Conflicts of Interest: The authors declared no conflict of interest.

Appendix A

To investigate $L(s)$, substitute $e^{-as} = \cos(\alpha w) - j\sin(\alpha\omega)$ into Equation (16),

$$L(j\omega) = \frac{j(k\omega + T\omega) + 1}{1 - \cos(\alpha\omega) + j[T\omega + \sin(\alpha\omega)]} \tag{A1}$$

Because of the effect of the trigonometric functions, the phase function and magnitude function of $L(j\omega)$ will oscillate in each cycle. Assume $\phi(\omega)$ is the phase of $L(j\omega)$, and we could get

$$\phi(\omega) = ac \ tan \ (k\omega + T\omega) - ac \ tan \ \frac{T\omega + \sin{(\alpha\omega)}}{1 - \cos{(\alpha\omega)}} \tag{A2}$$

If $\alpha = 0$, we have $\phi(\omega) = ac \ tan \ (k\omega + T\omega) - 0.5\pi < 0$. Hence, T should be big enough to make the loss of phase small at ω_c. Nevertheless, if T is too big, it will lead to the decreasing of the promotion for the system performance. Concentrating on the derivative function of $\phi(\omega)$ as follows,

$$\phi'(\omega) = \frac{k+T}{1+(k+T)^2\omega^2} - \frac{(T-\alpha)[1-\cos{(\alpha\omega)}] - T\alpha\omega\sin{(\alpha\omega)}}{[1-\cos{(\alpha\omega)}]^2 + [T\omega + \sin{(\alpha\omega)}]^2} \tag{A3}$$

Obviously, if $T < \alpha$, then $\phi'(\omega) > 0$ when $T\omega \in (0, \frac{\pi}{4})$. However, since α (the CCD's delay) cannot be too small, commonly, T has to be much bigger than α or it will result in instability. Actually, when $T > \alpha$ and $T\omega \in (0, \frac{\pi}{4})$, we can still get $\phi'(\omega) > 0$.

Proof.
$T\omega \in (0, \frac{\pi}{4})$.
Define $\psi(\omega) = (T - \alpha)[1 - \cos{(\alpha\omega)}] - T\alpha\omega\sin{(\alpha\omega)}$.
If $\psi(\omega) < 0$, obviously, $\phi'(\omega) > 0$.
Since $\psi(0) = 0$ and $\psi'(\omega) = -\alpha^2[\sin{(\alpha\omega)} + T\omega\cos{(\alpha\omega)}] < 0$, we get $\psi(\omega) < 0$. Then, $\phi'(\omega) > 0$.
From the above discussion, $\phi(\omega)$ is always increasing when ω is from 0 to $\frac{\pi}{4\tau}$. Note that

$$\begin{cases} \phi(0) = -0.5\pi \\ \phi(\omega_c) = ac \ tan \ (1 + T\omega_c) - ac \ tan \ \frac{T\omega_c + \sqrt{2}/2}{1-\sqrt{2}/2} < 0 \end{cases} \tag{A4}$$

It means that no matter what T is, the phase loss is inevitable. If we consider the amplitude of $L(j\omega)$, the same result will be obtained. The gain function of $L(j\omega)$ is as follows.

$$-20\log|L(j\omega)| = -10\log\left[\frac{(k\omega + T\omega)^2 + 1}{[1 - \cos{(\alpha\omega)}]^2 + [T\omega + \sin{(\alpha\omega)}]^2}\right] \tag{A5}$$

It is obvious that $-20\log|L(j\infty)| = -20\log{(\frac{k}{T} + 1)}$, $T >> k$ is hoped for reducing the sacrificing of GM. But,

$$-20\log|L(j\omega_g)| = -10\log\left[1 + \frac{4T\omega_g + 1}{4 + T^2\omega_g^2}\right] < 0 \tag{A6}$$

From Equation (A6), it can be concluded that the GM will also suffer losses. $\square$

References

1. Cochran, R.W.; Vassar, R.H. Fast-Steering Mirrors in Optical Control Systems. In *Advances in Optical Structure Systems, Proceedings of the SPIE 1303, Orlando, FL, USA, 16–20 April 1990*; SPIE: Bellingham, WA, USA, 1990.
2. Shlomi, A.; Kopeika, N.S. Vibration noise control in laser satellite communication. *Proc. SPIE* **2001**, *4365*, 188–194.
3. Liu, W.; Yao, K.; Huang, D.; Lin, X.; Wang, L.; Lv, Y. Performance evaluation of coherent free space optical communications with a double-stage fast-steering-mirror adaptive optics system depending on the Greenwood frequency. *Opt. Express* **2016**, *24*, 13288–13302. [CrossRef] [PubMed]
4. Leven, W.F.; Lanterman, A.D. Unscented kalman filters for multiple target tracking with symmetric measurement equations. *IEEE Trans. Autom. Control* **2009**, *54*, 370–375. [CrossRef]

5. Cao, Y.; Wang, G.; Yan, D.; Zhao, Z. Two algorithms for the detection and tracking of moving vehicle targets in aerial infrared image sequences. *Remote Sens.* **2015**, *8*, 28. [CrossRef]

6. Ekstrand, B. Tracking filters and models for seeker applications. *IEEE Trans. Aerosp. Electron. Syst.* **2001**, *37*, 965–977. [CrossRef]

7. Tang, T.; Huang, Y.; Liu, S. Acceleration feedback of a CCD-based tracking loop for fast steering mirror. *Opt. Eng.* **2009**, *48*, 510–520.

8. Yeung, K.S.; Wong, W.T. Root-locus plot of systems with time delay. *Electron. Lett.* **1982**, *18*, 480–481. [CrossRef]

9. Engelborghs, K.; Dambrine, M.; Roose, D. Limitations of a class of stabilization methods for delay systems. *IEEE Trans. Autom. Control* **2001**, *46*, 336–339. [CrossRef]

10. Srivastava, S.; Pandit, V.S. A PI/PID controller for time delay systems with desired closed loop time response and guaranteed gain and phase margins. *J. Process Control* **2016**, *37*, 70–77. [CrossRef]

11. Tang, T.; Ma, J.; Ge, R. PID-I controller of charge coupled device-based tracking loop for fast-steering mirror. *Opt. Eng.* **2011**, *50*, 043002. [CrossRef]

12. Cao, Z.; Chen, J.; Deng, C.; Mao, Y.; Li, Z. Improved Smith predictor control for fast steering mirror system. *IOP Conf. Ser. Earth Environ. Sci.* **2017**, *69*, 012085. [CrossRef]

13. Ren, W.; Luo, Y.; He, Q.N.; Zhou, X.; Deng, C.; Mao, Y.; Ren, G. Stabilization Control of Electro-Optical Tracking System with Fiber-Optic Gyroscope Based on Modified Smith Predictor Control Scheme. *IEEE Sens. J.* **2018**, *18*, 8172–8178. [CrossRef]

14. Jing, T.; Yang, W.; Peng, Z.; Tao, T.; Li, Z. Application of MEMS Accelerometers and Gyroscopes in Fast Steering Mirror Control Systems. *Sensors* **2016**, *16*, 440.

15. Dickson, W.C.; Yee, T.K.; Coward, J.F.; Mcclaren, A.; Pechner, D.A. *Compact Fiber Optic Gyroscopes for Platform Stabilization*; SPIE: San Diego, CA, USA, 2013; pp. 7453–7458.

16. Yoon, Y.G.; Lee, S.M.; Kim, J.H. Implementation of a Low-cost Fiber Optic Gyroscope for a Line-of-Sight Stabilization System. *J. Inst. Control* **2015**, *21*, 168–172.

17. Tang, T.; Huang, Y. Combined line-of-sight error and angular position to generate feedforward control for a charge-coupled device–based tracking loop. *Opt. Eng.* **2015**, *54*, 105107. [CrossRef]

18. Xie, R.; Zhang, T.; Li, J.; Dai, M. Compensating Unknown Time-Varying Delay in Opto-Electronic Platform Tracking Servo System. *Sensors* **2017**, *17*, 1071. [CrossRef] [PubMed]

19. Böhm, M.; Pott, J.U.; Kürster, M.; Sawodny, O.; Defrère, D.; Hinz, P. Delay Compensation for Real Time Disturbance Estimation at Extremely Large Telescopes. *IEEE Trans. Control Syst. Technol.* **2016**, *25*, 1384–1393. [CrossRef]

20. Glück, M.; Pott, J.U.; Sawodny, O. Piezo-actuated vibration disturbance mirror for investigating accelerometer-based tip-tilt reconstruction in large telescopes. *IFAC-PapersOnLine* **2016**, *49*, 361–366. [CrossRef]

21. Kim, B.K.; Wan, K.C. Advanced disturbance observer design for mechanical positioning systems. *IEEE Trans. Ind. Electron.* **2004**, *50*, 1207–1216.

22. Luo, Y.; Huang, Y.; Deng, C.; Yao, M.; Ren, W.; Wu, Q. Combining a disturbance observer with triple-loop control based on MEMS accelerometers for line-of-sight stabilization. *Sensors* **2017**, *17*, 2648. [CrossRef] [PubMed]

MDPI

St. Alban-Anlage 66

4052 Basel

Switzerland

Tel. +41 61 683 77 34

Fax +41 61 302 89 18

www.mdpi.com

Electronics Editorial Office

E-mail: electronics@mdpi.com

www.mdpi.com/journal/electronics